Linguistische
Arbeiten 177

Herausgegeben von Hans Altmann, Herbert E. Brekle, Hans Jürgen Heringer,
Christian Rohrer, Heinz Vater und Otmar Werner

Process Linguistics

Exploring the processual aspects of language and language use, and the methods of their description

Edited by
Thomas T Ballmer † and Wolfgang Wildgen

Max Niemeyer Verlag
Tübingen 1987

CIP-Kurztitelaufnahme der Deutschen Bibliothek

Process linguistics : exploring the processual aspects of language and language use, and the methods of their description / ed. by Thomas T. Ballmer and Wolfgang Wildgen. – Tübingen : Niemeyer, 1987.
(Linguistische Arbeiten ; 177)
NE: Ballmer, Thomas T. [Hrsg.]; GT

ISBN 3-484-30177-5 ISSN 0344-6727

Printed in Germany. Druck: Weihert-Druck GmbH, Darmstadt.

TABLE OF CONTENTS

INTRODUCTION

Wolfgang Wildgen
University of Bremen

The volume "Process Linguistics" contains the second series of articles collected by Thomas Ballmer in the years 1981 and 1982. He had the intention to represent the entire range of scientific activities in this new sector of linguistic research. The first series of papers was completed when Thomas Ballmer died (2nd december 1984) and my colleagues Hannes Rieser and Hans-Jürgen Eikmeyer took care of it. The volume "Linguistic Dynamics. Discourses, Procedures and Evolution" appeared in the collection "Research in Text Theory" (De Gruyter, Berlin, 1985). The original introduction to the two volumes written by Thomas Ballmer is published as a section of this book (pp. 1-14).

As the second volume was not complete and finished, it had to be modified. I included some new articles: Thomas Ballmer's essay: "Case, Aktionsart and Ergativity" and an article by myself, which is complementary to Ballmer's paper and elaborates some ideas put forward by him. Some of the articles originally submitted had been published in the meantime, some have been withdrawn or revised by the authors. The general character of the volume, however, is in accordance with Ballmer's original concept.

Almost five years after Thomas Ballmer's initiative we can look back on the "dynamic movement", which began in the mid-seventies and had its heyday between 1981 and 1985. As a point of departure of our "historical" reflection we can consider the first announcement of the volume "Linguistic Dynamics", written by Thomas Ballmer in 1981.

During the last few years a view of *linguistic* phenomena was given a new impulse and revival which is best characterized as *dynamic*.
In contrast to an exclusively relational and static conception, the dynamic view concentrates on linguistic processes. Temporality is in focus. The production, flow and interpretation of utterances and thoughts are considered in temporal detail. Procedural mechanisms of language are discussed. The creation and processing of discourse are of increasing concern: motivation, goals, intention, planning, action, context change, performance, reactions. In addition the mental and linguistic procedures on the level of speech synthesis and analysis, the dynamic viewpoint of language development, of historical and maybe even of biological evolution become of interest. Psycholinguistic, sociolinguistic and anthropological questions are raised as well as questions of second language learning and teaching. Theoretical strands of linguistics - logical grammar, artificial intelligence, theory of grammar, etc. - as well as empirical and applied fields of linguistics can make use of the new perspective of analysis.

There is the idea of collecting original papers focusing upon the various topics of *linguistic dynamics* - as this newly developing viewpoint may be called perhaps - and to publish them in a volume.

The background of this enterprise were scientific activities and publications since 1978, especially the discussion in the Bielefeld working group on "Language and Logics" (since 1974). In 1978 a workshop was held which discussed the problem of *vagueness* and *fuzziness* (published in Ballmer and Pinkal, 1983). Some of the work was also documented in the volume: "Empirical Semantics I, II" edited by Burghard Rieger and in the reader: "Words, Worlds and Contexts. New Approaches in Word Semantics" edited by Hans-Jürgen Eikmeyer and Hannes Rieser. The group of young scholars interested in the dynamical and procedural aspects of language also met at several congresses such as:

- The Annual Meeting of the German Society of Linguistics (DGfS) in Regensburg (1981). Thomas Ballmer organized a workshop on "Dynamic and procedural aspects of semantics".

- The Semiotic Colloquium at Hamburg (1981), where H.-J. Eikmeyer and H. Rieser directed a workshop on "The constitution of meaning".
- The Annual Meeting of the DGfS in Cologne (1982), where Thomas Ballmer chaired a workshop on "Dynamic and processual aspects of semantics".
- The German Meeting of Germanists in Aachen (1982) where Burghard Rieger organized a workshop, the papers where published under the title "Dynamics in the constitution of meaning".
- The workshop "Procedural Linguistics" in Bielefeld organized by H.-J. Eikmeyer and H. Rieser (1982).
- The Roundtable: "Dynamical Aspects of Language and Logics" organized by Arnold Oberschelp et alii in Kiel (1983). Meetings dealing with more specific topics at Passau (1983) and Bielefeld (1984). At the Bielefeld Meeting of the German Society of Linguistics Herman Haken delivered the inaugural speech: "Is language a synergetic system?". The co-referents were Thomas Ballmer, Hans-Jürgen Eikmeyer and Wolfgang Wildgen.

The group of dynamicists who met on these occasions had very different scientific interests and applied different mothods. They were generally interested in methodological and theoretical innovations which could integrate the multiple advances in the interdisciplinary study of language without losing the high level of formal and mathematical sophistication reached in theoretical linguistics. We can distinguish several centres of interest:

- textual and conversational processes,
- processes of language use, linguistic innovation and language change,
- morphogenetic and evolutionary aspects of language,
- computational models of language processing (procedural models),
- applications of the theory of (nonlinear) dynamic systems

in linguistics (catastrophe theory, bifurcation theory, theory of self-organization, synergetics etc.).

The death of Thomas Ballmer, who was interested in all these fields hastened the separation into different subgroups, each of which tried to find a stable methodological frame and to consolidate first results in empirical research. A unified theory of linguistic dynamics became only a long-term goal. In a sense this and the preceding volume "Linguistic Dynamics" document a first stage in the scientific development. A subsequent stage will lead to divergence *and* consolidation.

The papers contained in this reader are ordered into three groups.

The first group of papers explores the consequences of a radical shift of paradigm in theoretical linguistics: from the discrete, structural, combinatorial paradigm to the continuous, dynamical paradigm. The term "dynamics" gets a precise reading founded in the mathematical theory of dynamic (especially non-linear) systems. In this narrow sense a dynamic theory of language or a dynamic model is a theory or a model formulated within the framework of the kind of mathematics developed in catastrophe theory, bifurcation theory and nonlinear systems theory (cf. Casti 1985). The first paper in this group sketches the theoretical programme of the continuous, analogue and dynamic modelling of linguistic phenomena. *Tim Poston*, a specialist in mathematical systems theory, shows some interesting applications of the new paradigm to the phenomenon of meaning shift and to the geometry of lexical fields. The second paper by *Thomas Ballmer* sets out with a summary of results obtained in the analysis of the German lexicon of verbs (cf. Ballmer and Brennenstuhl, 1985). The central notion of a three-dimensional semantic space is applied to the description of case, ergativity and 'Aktionsart'. As for the theoretical background, he refers to bifurcation theory and to the theory of self-organizing systems. The lexicological perspective is taken up in the paper by *Wolfgang Wildgen*, who summarizes results of catastrophe theoretic semantics and considers basic fields of lexical categorization, in the field of basic colour terms, of evaluative adjectives and of verbs for basic actions). In its last section the paper elaborates Ballmer's concept of a semantic space within the framework of catastrophe theoretic semantics. A second paper by *Wolfgang Wildgen* applies catastrophe theoretic schemata to the description of nominal compounds in German. He arrives at some basic dynamic principles which govern the coining and the interpretation of nominal compounds in German.

By contrast the second group of papers is concerned with empirical questions related to dynamic aspects of language. *Walter Kindt* asks the fundamental question: "why have linguists not so far freed themselves of such static views and applied themselves to the development of dynamic conceptions of language?" (this volume, pp.167f). He argues in favour of a more empirically oriented semantics: "without stronger empiricism of semantics it will not be possible to develop dynamic models; without dynamic modellings, however, no effective semantic theories applicable to practical problems will be obtained". (ibidem, p. 168).

The papers which follow try to approach processual aspects of language in a rather informal way. *Ton van der Geest* gives a comprehensive picture of language acquisition and language learning and evaluates the controversy between the (Chomskyan) language acquisition theory (LA) and the psychological theory of language learning (LL). *Florian Coulmas* deals with the dynamics of language use and of actual changes in language use. The creation of new derivations, compounds, metaphors, idiomatic expressions (in English) allows for an empirical analysis of actual processes of linguistic innovation. His results encourage the reassessment of basic issues in linguistic theory such as: the arbitrariness and complexity of the linguistic sign. *Marina Sbiza* presents the current discussion on speech act theory and develops a dynamic model of the illocutionary act. Speech act theory "ought to be the study of the changes that speech acts bring about in their own contexts" (in this volume, p. 252) and should finally" succeed in developing into a theory of human interaction" (ibidem, p. 278).

In the last group of papers questions of language processes are treated within the framework of traditional formal or technical methods. *Alice ter Meulen* considers discourse phenomena within the framework of modeltheoretic semantics. Her paper marks a specific stage in the development of these models, i.e.: "the transition from Montague Grammar with its complete information models to dynamic interpretation based on partial information models" this volume, p. 383). In *Anthony Jameson's* and *Wolfgang Wahlster's* paper a computerized dialogue system (HAM-ANS) constitutes the background for the simulation of the generation of anaphora, ellipsis and definite descriptions. Their system is able "to take into account the existential assumptions, domain-related desires, and referential beliefs ascribed to the partners." (this volume, p. 309). In the contribution by *Ju. K. Orlov* the flow of speech is analysed statistically. He discusses the results of structural and descriptive models following Herdan's probabilistic concept of language, with particular reference to current research done in the URSS.

The studies brought together in this reader reveal the large spectrum of processual and dynamic models and descriptions. In terms of continuous scales we could arrange them between two extremes:

A. Mathematical models for continuous, nonlinear processes (with self-organization or stabilization) applied to the field of linguistics.

B. Conservative structural models augmented by techniques for the covering of "dynamical" aspects (logical models with "dynamic" interpretation, probabilistic models for the lexicon and computational models with *procedures* instead of *structures*).

In the middle section of the scale we find empirical enterprises which concentrate on dynamical aspects of language, language use and the acquisition, learning, innovation and change of language. At present it is necessary to develop the whole field, although dynamic system theory promises to become the backbone of the dynamic and processual paradigm, as it opens up the way to a large interdisciplinary field from neurodynamics (cf. Scott, 1977) to the social dynamics of interacting populations (cf. Weidlich and Haag, 1983). A first integrated view of dynamic linguistics based on dynamic systems theory is unfolded in Wildgen and Mottron (forthcoming 1987).

References

Ballmer, T.T (ed.), 1985, *Linguistic Dynamics*. Discourses, Procedures and Evolution. De Gruyter, Berlin.

Ballmer, T.T and W. Brennenstuhl, 1985. *Deutsche Verben*. Narr, Tübingen.

Ballmer, T.T and W. Kindt (eds.), 1980. *Zum Thema Sprache und Logik*. Ergebnisse einer interdisziplinären Diskussion. Buske, Hamburg.

Ballmer, T.T and M. Pinkal (eds.), 1983. *Approaching Vagueness*. North Holland, Amsterdam.

Casti, J.L., 1985. *Nonlinear System Theory*. Academic Press, London.

Eikmeyer, H.-J. and H. Rieser (eds.), 1981, *Words, Worlds, and Context*. New Approaches to Words Semantics. De Gruyter, Berlin.

Rieger, B. (ed.), 1981, *Empirical Semantics*, Vol. I, II. Brockmeyer, Bochum.

Rieger, B. (ed.), 1985, *Dynamik in der Bedeutungskonstitution*. Buske, Hamburg.

Scott, A.C., 1977, *Neurodynamics*. A Critical Survey. In: *Journal of Mathematical Psychology*, 15, pp. 1-45.

Weidlich, W. and G. Haag, 1983. *Concepts and Models of a Quantitative Sociology*. The Dynamics of Interacting Populations. Springer, Berlin.

Wildgen, W. and L. Mottron, (forthcoming 1987), *Dynamische Sprachtheorie*. Brockmeyer, Bochum.

PART ONE: MODELLING PROCESSUAL ASPECTS OF LANGUAGE IN THE FRAMEWORK OF (NONLINEAR) DYNAMIC SYSTEMS THEORY

"MISTER! YOUR BACK WHEEL'S GOING ROUND!"

Tim Poston
University of California, Santa Cruz

Abstract

Human speech signals have many analogue aspects (pitch, rhythm, volume ...), processed before other features by the hearer, which condition the semantic and syntactic processing of the embedded strings of words. To treat words as discrete signals, as the 'text' concept of speech acts suggests, leads to a combinatorial model of the way they are/should be processed; that is, of their meaning. But meanings can change continuously over time, contrary to combinatorics, and when they do shift discontinuously the jumps may be better modelled by continuous dynamics with bifurcating attractors than by discrete models. In an example, meaning reversal is related to the cusp catastrophe (ubiquitous in nonlinear dynamics and alien to combinatorics) with the possibility of experimental test. Its confirmation would imply the need for a radically different approach to the construction of grammars.

Keywords

analogue, bifurcation, bistable perception, catastrophe theory, disambiguation, frame, hysteresis, linguistics, nonlinear dynamics, speech recognition

1. My title is taken from a speech cognition experiment beloved of my schoolmates. Shout it in the right tone to a cyclist and he will respond - fast - to the prehuman alarm signal of the urgent call, and skid to a halt. As processing of the verbal content proceeds, he looks at his rear wheel; and when it is complete, he laughs or scowls according to temperament.

This example illustrates that human processing of a string of words is conditioned strongly by their volume, pitch and rhythm, which are analysed faster. The message is long completed, and the bicycle stopped, at a point where the brain is still trying to analyse the signalled danger to the wheel, and consulting its sensory apparatus (mainly vision) in the attempt. The question 'What's wrong with the wheel?', which at whatever level of articulation is clearly involved in this activity of analysis, would never have arisen if the tone of the call had been (for instance) admiring. The cyclist would have ridden steadily on, and decided after analysis that the compliment paid his machine was an irritatingly empty one. Danger would not have been part of the 'frame' (to use a word fashionable in machine linguistics) that governed the work of interpretation.

This significance of pitch and rhythm in controlling the frame for analysis is largely neglected in the literature. No adequate notation exists for it; phonetics texts record the rising tone that signals a question in many languages, and so on, but what carries the message "I am telling a joke."? How do I know, sometimes, that a person I telephone is smiling? These key indicators are left untranscribed. And yet, an excerpt of conversation is commonly referred to as a *text*, and a transcript treated as equivalent to a recording. Of course, since no machine can yet produce an adequate transcript of fluent speech, a true text holds more machine-usable linguistic information than does a recording. But humans are not IBM products, and it is almost as hard to pronounce 'Fruit flies like a banana' ambiguously as to confuse 'lead' the verb and 'lead' the metal. Syntax and semantics are voice-dependent. It does not need the linguistic incompetence of a Nixon (though it helps) to produce transcripts nearly unintelligible to a human reader. This result is common anyway, with speech that when heard is perfectly clear. No account I have read, or heard, of machine analysis of conversation has convinced me that the raw material fed in - let alone the programming - was adequate for the job. (I recall one elaborate story-telling analysis, by a researcher imitating the machine, which entirely missed the point that the storyteller intended to be funny. The fact was audibly blatant, and organised the interplay between the storyteller and his wife, but the researcher's machine imitation had reached the level of hearing only the transcribable. This is nearer autism than objectivity.) To hope that machines will find intelligible what human users of language do not is logically possible, but surely rather more optimistic than current programs (as against claims) can support.

As an aside, it is worth noting that not only does a novelist need to write conversation unlike any transcript, to achieve intelligibility; the difference between a live-seeming fictional dialogue and a dead one is that the good writer gives successful signals, without being clumsy or blatant, of the tone in which the reader should hear the things said. The speech can then be stored in the reader's memory as though it were real speech.

2. Analogue versus digital description

The implication of the remarks above is not merely that conversation analysts should enlarge their data set (with a somehow improved transcription code) and proceed as before, while workers with texts in the true sense (such as translators of the written word) have not even that modification to make. It is that human language, in its original vocal use (both historically and with individuals, far prior to reading and writing) is enmeshed in a system which is deeply continuous, or analogue[1], in its character,

---

1 'Analogue' here comes from computer usage; an analogue computer has (by arrangement) dynamics analogous to those of the system modelled. The fact that it has become almost a synonym for continuous as against digital description or simulation is an implicit acknowledgement that real systems are almost always continua, so that the analogy can hold in one case but not the other (is a finite-state automaton analogous to any naturally occurring system?), but this acknowledgement has largely been suppressed by the 'Artificial Intelligence' community. 'Analogue' as 'continuum theoretic' here is thus wider than 'analogic' versus 'digital' in (Bateson, 1966), where 'digital' overlaps with *conventional* assignment of meaning, as in graphical versus alphanumeric symbols. The different pitch structures of "What?" in

(continued on page 14)

not discrete, digital or combinatorial like (for instance) Chomsky's transformational grammar. The structural consequences are profound.

Any claim for such consequences rests on the existence of a real analogue/digital contrast, which is often casually denied. Every continuous system can be digitally approximated, in some sense, arbitrarily well. But in what sense? Digital recordings are, of course, superb, but no text represents speech by remotely so many bits. (A good description using 10^{5+}bits/sec. does not imply that a textual record at 20 letters/sec. will also be an adequate account.) Furthermore the approximation of dynamics is a far more subtle problem than is often suggested in accounts of numerical analysis. There are many continuous models, stable and successful in physical prediction, which are remarkably hard to put on a computer. The natural finite difference schemes blow up after a couple of thousand steps, and the numerical analysts ("the academic world's greatest plumbers"; Briggs et.al 1981) patch the structure with some *ad hoc* artificiality, to survive for a thousand or two more steps. Shrinking the step size generally helps, though in numerical weather prediction (for example) it makes things worse by finding sound waves as well as bulk motion, so that a fix goes the other way. Anyway, steps cannot be shrunk

---

(continued from page 13)

"It's coming!"		"It's coming!"
"What?"	and	"What?"
"I said, it's coming!"		"The train!"

are not prehuman linguistic universals, like an angry tone of voice, let alone structurally analogous to some referent, but they are continuously variable and hence analogue signals in the sense used here.

ad infinitum; and theorems describing the convergence of difference schemes to the solutions of nonlinear continuous models are thin on the ground. Theoretical 'discrete dynamics', more closely related to differential equations, discretises time but not the state variables.

Moreover, though the picture of the brain as a computer-like assemblage of finite-state neurones is popular among computer specialists, it has little support from neurology. Neurones are discrete units, but their firing rates, refractory periods, ionic permeabilities etc. are continuous variables, whose highly nonlinear interactions remain ill understood. Modelling them by two-state switches may (or may not) produce useful results for a particular purpose one has; but it has no general, reductionistic guarantee. An analogue view of the brain is at least as defensible, and where analogue and digital views - at a given level of detail in description - produce conflicting expectations of its behaviour, only experiment can judge between them. A tentative model whose predictions are out of character for finite state theories of language is described in Section 5, below.

3. Continuous models of language

I return then to language, hopefully with at least provisional acceptance by the reader of continuous dynamical terminology in describing the brain's processing of speech. By implication such language then enters the description of language itself; for taking the view that the meaning of a word consists at least in part of its potential to affect subsequent behaviour (verbal or otherwise), the meaning is then factored through an analogue interpretation process. To put it another way, the dynamics of the human brain is a part of English and Chinese, just as the discrete dynamics of the digital computer is a part of FORTRAN IV and LISP. Meaning is not a Platonic something a word has, it is a Buddhist something the five skandhas do with it.

What, then, might a continuous dynamical description of language involve, and how would it differ operationally from a combinatorial one? The following remarks are not claimed to constitute a theory, but do provide one possible frame for discussion and lead to suggestions that should be testable.

A point in the 'phase space' of the brain is specified by as many numbers as it takes to give the state of each of its $10^{\text{roughly}\,10}$ neurones. The paradigmatic theory of the axon (the Hodgkin-Huxley model) requires four real variables, so it seems unlikely that one can describe a whole neurone with less; let us say, between 10 and 1000 milliard real numbers to specify a brain state. These numbers are too vast for the odd order of magnitude to concern us more than it does an astronomer, but fortunately in non-linear dynamics there are very often conditions satisfied that allow a strict reduction of a problem to just a few variables. There are some partial explanations of why this is so often true, but since in some circumstances infinity is a better approximation to 10^{10} than 10 is (Magnus and Poston 1979), and the relevant theorems guarantee the reduction only over some range, perhaps smaller than can be measured, only mystification would be served by enlarging on them here. Their language is impressive, but their content is in practice only that the success in physics, engineering and chemistry of relating measured quantities to theoretical reduced variables seems reasonable. (The hard sciences have fewer 'exact solutions' than their public images suggest, and these are almost always solutions of highly approximate, idealised models.) However, one point is

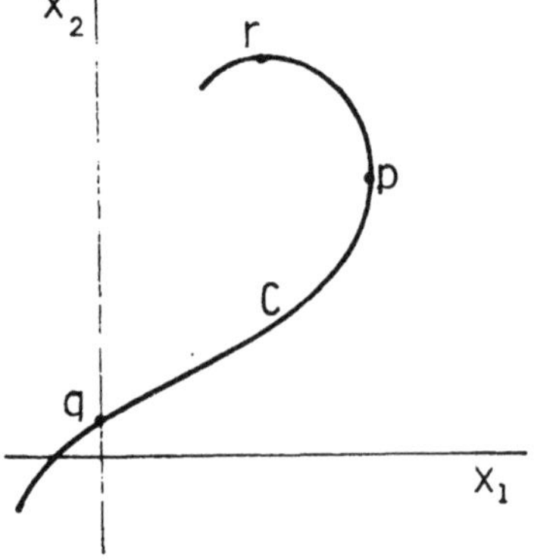

Fig. 1 Near q, points on C correspond one-to-one with their x_1-coordinates, and also with their x_2-coordinates.

experimentally relevant. If for instance a system described by two variables x_1 and x_2 (Figure 1) is confined for the interesting part of the dynamics to a curve C, returning to it almost instantly if pushed off, then either x_1 *or* x_2 will do to describe its position in C - the 'interesting part' of its state - in the region around q. Past p, x_1 fails, and past r, x_2 fails, but near q they both work. In a sense this is typically true around most points (see Poston and Stewart 1978a, p. 414, for more discussion). Indeed with any number of variables x_1, x_2, ..., x_N, not merely the x_i but *anything we can measure,* that is coupled reasonably with our phenomenon of interest and not too strongly with uncontrollables like yesterday's breakfast, will usually do as a coordinate on C and its higher-dimensional analogues.

The net result is that any phenomenon we can describe with a few variables can be imitated (and in the hard sciences often is) by a system that has a great many. Low-dimensional suggestions like those below of things that may happen are not *ipso facto* unrealistic, though one should not make a general claim that only low-dimensional things *can* happen. (There is a certain limited sense in which such a claim can be proved for 'elementary catastrophe theory', but it is certainly false for more general dynamics. For an inspired non-technical introduction to the phenomenology of non-linear dynamics, see (Abraham and Shaw 1982).)

Suppose then that we wish to describe the process of decision as to whether 'your ass', for example, refers to your donkey or your buttocks. (Modern British English has a phonetic distinction which shows up in spelling; the languages of Tristram Shandy, gentleman, and the United States of America do not.) A natural approach, dynamically, is to think of each final decision as an 'attractor' like points D and B in Figure 2.

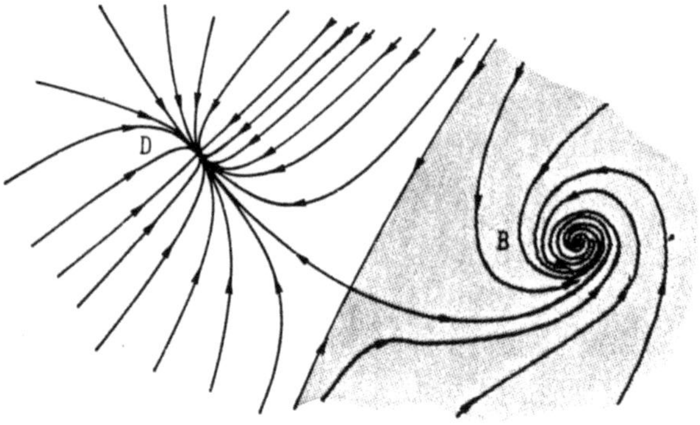

Fig. 2 A flow (ordinary differential equation) on the plane, with two attracting points and a saddle.

Tone, context etc. between them fix the starting point and the dynamical rules (arrows) to be followed. (The fact that one is talking to a bishop may act more like a fixed parameter affecting the dynamics, so that D catches more points, than an initial value. But this distinction is more descriptive than fundamental.) If the starting point is in the shaded region, the decision goes to B; if in the unshaded, to D.

Now, in Figure 2 the distinction between the 'basins of attraction' of (sets of point drawn to) D and B is near a linear one, decidable by even such a primitive artificial intelligence as the perceptrons whose limitations are so clearly analysed in Minsky and Papert (1965). But in Figure 3 the basins of B (dark shaded) and D (unshaded) are wrapped infinitely many times around a repelling closed loop (*i.e.* one from which nearby points depart). The curve separating them has become so intricate that no simple algorithm will decide which side of it a point lies on, short of 'run the system [or a simulation of it] and see where it ends up'.

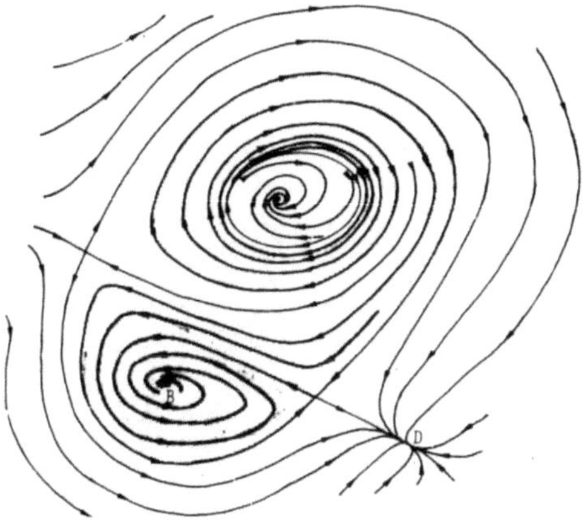

Fig. 3 A flow with three attracting points, two saddles, and a repelling cycle.

Every point on the edge of the basin of attraction (light shaded) of the third attractor is also in the edge of both the other basins. We could equally have wound any number of basins together on the outside of the cycle, and any other number on the inside, without leaving the class of stable-in-shape-under-small-perturbations flows in two dimensions. Three dimensions, or two with discrete time, give examples with simple formulae where whenever a point is on the boundary of two basins, it is also on the boundary of a third.

Even if we assume the dynamics to be of gradient type (not a reasonable restriction for the brain), which forbids such fine tangling of basins, we can have examples like Figure 4, in which one basin entirely surrounds another.

In general, even with the simplest non-linear systems the choice of an attractor is topological and dynamic, not a metrical 'find the nearest' scheme like the principle universal in (for instance) computer speech recognition routines, which can generate nothing subtler than Figure 5.

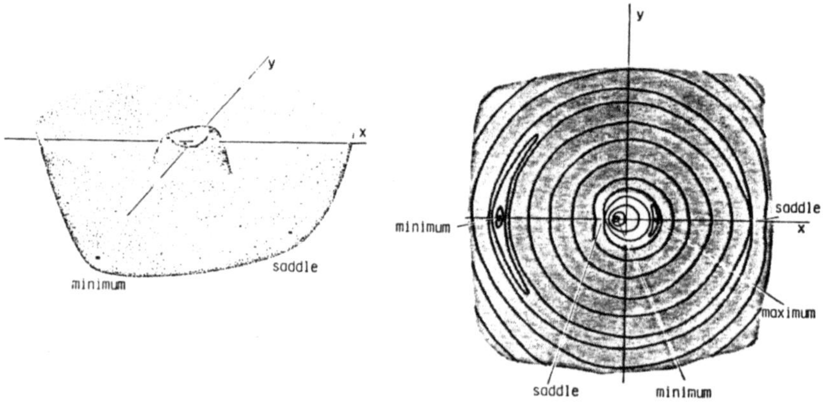

Fig. 4 Graph and contour map of $2(x^2+y^2)^3-6(x^2+y^2)^2+x^2+y^2+\frac{1}{6}x$. The basin of attraction (shaded) of the deeper minimum, under the gradient flow, completely surrounds the basin of the other.

It is hardly surprising that computer accuracy in speech recognition is dismal; when a metric scheme says that a sound recognised as 'the' by any English speaker is 0.314 distant from its 'the' template, while only 0.242 from 'seven' (Levinson and Liberman 1981), it is probably not the choice of metric but the metric

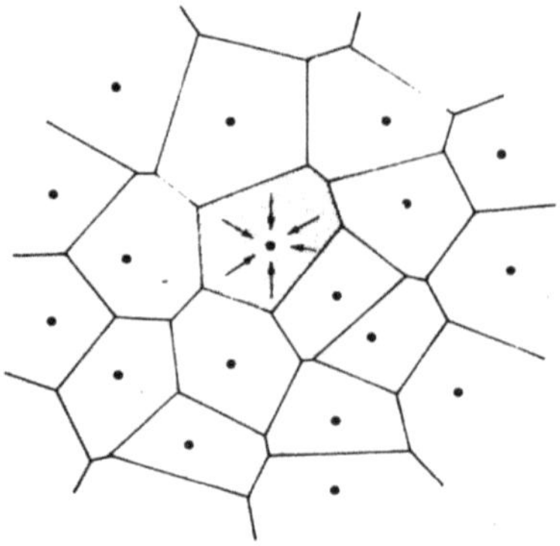

Fig. 5 Basins of attraction of the points marked, under the 'go to the nearest dot' dynamic assumed in all nearest-template-finding computer matching schemes.

approach that is wrong. Modelling what the brain does is difficult, but probably for all speech sciences unavoidable. In this context one could begin by mapping out what machine-generated sounds are recognised by humans as what words, and how fast the assignment is in each case, as a step towards the dynamic. One could at least decide whether the topology is as subtle as Figure 3, or as relatively simple as Figure 4.

In higher dimensions the topological relationships between attractors can be far more complicated; in computer 'association network' models of language, just as in speech recognition, the relations between words are assumed far simpler.

The claim here, of course, is not that continuous dynamics are more complicated and subtle than combinatorial algorithms; Church's Thesis suggests that nothing with clear results *can* be more subtle. It is that to imitate continuous systems - even low-dimensional ones given by simple polynomials - with discrete ones can require a fine mesh and small steps to get the imitation qualitatively plausible. (It is noteworthy that computer investigations of 'chaos' in simple-looking dynamical equations often need to work with multiple precision.) Linguistically, this fine mesh would correspond to recording fine details of pitch, rhythm, etc.; or, in a written text, allowing so many fine variations of a 'frame' that the intellectual concept becomes a *continuum* of frames, discretised by computational necessity, not a short finite list. Computing which frame was appropriate would involve far more data about the words (and their combinations) than is usually placed in a dictionary or program. Connotations of class, race, humorous intent, dignity, ..., can all radically change the interpretation of a sentence, to the point of reversing it entirely. "I speak not to disprove what Brutus spoke", in context, gives Brutus the lie - and with many recordings of the *spoken* words, the context could be omitted and the same meaning heard. To an extent, vocal rendering and context provide mutual redundancy, and thus at some level of analysis the same kind of information. Either the brain processes speech characteristics in some digitised form, or it treats written words as inputs to an evolving analogue (or partly analogue) entity we call 'context' or 'frame' according to ... context.

The main deduction, or 'prediction', so far from the suggested analogue character of the structure of language is that it is liable to be hard to reduce to coarse-grained word combinatorics. So far, spot on, as anyone who has worked on automatic text analysis will agree, but so what? Imitating an unknown computer program's response to input can be pretty tough, too. The continuous modelling suggestion will only be fruitful if it can describe in a reasonably simple way some observed phenomena which resist modelling by combinatorics. The next section suggests an area in which this may hold true, illustrated by an example in Section 5.

4. The phenomenon of meaning shift

The meaning of a word often drifts in time, by a gradual shift of connotations and associations. For example, by its very nature the mildly archaic flavour of 'to hasten' cannot have been acquired suddenly, unlike the totally 'unhip' character that 'hep' acquired in a fashion shift a couple of decades ago. Mild datedness comes only from a slow passage out of use. This flavour is part of its current meaning, since it biases the analysis of other words in the same speech act towards their older, or more high style, meanings. ('Hasten here with your ass!' attracts one of the meanings for 'ass' discussed above, 'Get your ass over here, fast!' the other.) Meanings, and the frames they contribute to, can evolve as continuous variables.

They can also jump violently, by sarcasm ("and Brutus is an honourable man"), metaphor ('This bus is a shuttle.'), and so on. 'Testa' in Latin meant 'chamberpot', and via low Latin gave us the French 'tête'. 'Nunnery' in Elizabethan slang meant 'brothel'. 'Nice' has leapt around, as we find in the OED, between 'foolish', 'wanton', 'elegant', 'strange, rare', 'slothful', 'effeminate', 'shy', 'fastidious', 'scrupulous', 'cultured', 'intricate', 'exact', 'thin', 'dainty, appetising', and 'agreeable' among other meanings. 'Fine' has mainly positive meanings, but can strengthen negative words. 'Presently' has switched from 'at once' to 'sometime soon', while the old sense 'at present' is presently returning to England from the USA. 'Gift' in German has changed from its English meaning to 'poison', though Old High German had both senses and 'Mitgift' still means 'dowry'. And so on.

At first blush these conspicuous discontinuities argue against a continuous model for meaning, and so - by implication - in favour of a combinatorial one. But it is exactly in studying them that the difficulties of a combinatorial approach become most acute.

When a word evolves gradually, we may suppose that the community of its users assigns 'a' meaning to it. Ignoring nuances of difference between generations and individuals, the coarse-grained approach of treating it as a combinatorial point may have non-trivial success. But when a word jumps in meaning, both meanings coexist for a while; such shifts are never originated by Duden or the Académie Française, with the linguistic community goose-stepping into line. The new sense originates (often in some subculture) and propagates, and where it is presently growing there are many speakers ready not only to take, but to use, the word in both the old and the new recension. The new sense may be 'slang' for a while, like 'testa' and the example in the next section, but I will ignore this often-temporary pejorative.

In such circumstances, how is the choice indicated by the user and made by the hearer? Once again, by tone, rhythm and context. (Hamlet to Ophelia, "Get thee to a nunnery, go!" was probably far less ambiguous, one way or the other, to a contemporary hearer than to a modern scholar; though *all* senses of a word can resonate in its choice by a poet.) Often, extremes of tone will disambiguate a word entirely. But do intermediate values of these untranscribed variables act so neatly? Is there a frontier, to one side of which one hears D, and on the other B (even a frontier as intricate as in Figure 3), or is there a range of these variables in which the decisive factor - if any - is context?

To process a word with the help of context cannot mean to reprocess the whole material of the last half-page, half-chapter or half-hour's conversation. (Recall the waiter who added a long list of prices for food and drink, and when someone remembered one more beer repeated the whole addition. To the suggestion that he add the one price to his previous total, for the same result, he said simply "I am not a philosopher".) The processing time, whether digital or analogue, would explode. Rather, the results of previous processing are carried as a 'moving frame', or evolving system of bias in the way words are treated; arguably, a system in *continuous* evolution for much of the time, though it too may sometimes jump. The meaning given to word depends on the recent history of words read or heard, via a system of continuous variables.

It is convenient here to draw an analogy with another disambiguation process where a meaning is to be chosen.

Fig. 6 An ambiguous figure due to Fisher (1967).

The drawing in Figure 6 was reported by Fisher (1967) to be perceived with equal probability as a man's face, or as a kneeling girl, when presented free of context.

Attneave (1971) embedded it in a series (Figure 7).

Fig. 7 Attneave's (1971) biasing sequence for Figure 6.

He found that subjects who had just seen the unambiguous 'man' at left saw the original figure that way with high probability, while those approaching in sequence from the right usually saw 'girl'. As we go from one end of the sequence to the other, at some point past the middle the concept assigned to the input jumps. (Freudian or other aspects could shift the bias for individual subjects. But the features of concern here, of unambiguous ends and an ambiguous central region where the decision depends on recent history, would only move a bit, not disappear.) Then, in (Poston and Stewart 1978b), the sequence was enlarged to a grid (Figure 8) around which

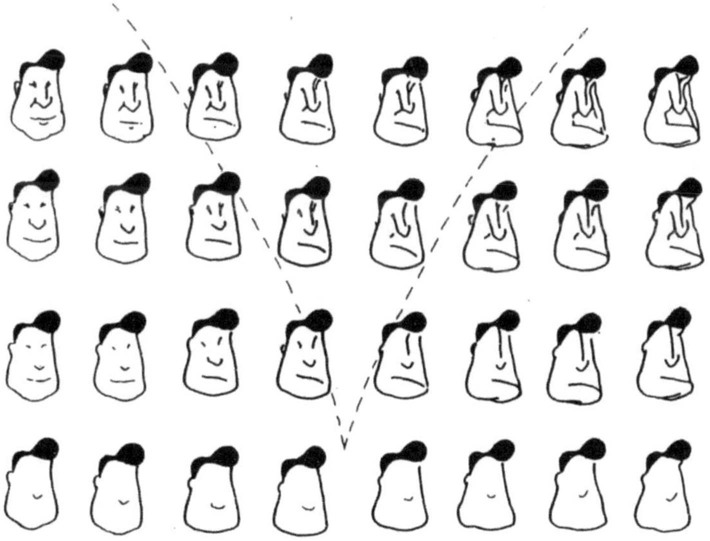

Fig. 8 The effect of loss of detail on Figure 7, from (Poston and Stewart 1978b).

a gradual change from seeing 'man' to seeing 'girl', with no experience of a sudden shift, is possible.

I will not repeat here the somewhat involved reasoning, some mathematically strict and some heuristic, that suggests the dashed curve as the natural form for the set of points where perceptual jumps take place as the figure is continuously changed (for example, by computer animation). It arises, however, out of an essentially analogue picture of the brain. The cusped geometry involved arises with monotonous regularity in the behaviour of continuous dynamics, and with more general 'attractors' than just point equilibria. For instance, in the Duffing equation for two coupled oscillators the attractors are periodic behaviours, and Zeeman (1976) discusses this as motivation for expecting the cusp sometimes to organise shifts in the brain's behaviour. At a more primitive level, it has been shown (Harth *at al.* 1975, Cowan and Ermentrout 1978) to organise dynamics of some phenomena in neural nets, such as the escape mechanism of *Tritonia*.

It has no combinatorial reason to exist.

It can easily be approximated combinatorially, though often associated with 'stiff' equations which give numerical trouble requiring many small calculation steps for reasonable precision. But it shows up approximated in discrete models only when these are designed as approximations of continuous ones. Its characteristic features, such as the biconcavity of the cusped curve, are difficult even to describe combinatorially, let alone to derive as one does for a broad class of continuous dynamics. Its ubiquity is natural in a continuous, differentiable setting (though only in a limited context within elementary catastrophe theory can one show that nothing *else* should be expected), while it has no natural link with finite-state automata. (Precisely because it is not ubiquitous for *all* systems, its occurrence when observed carries information about the class of system studied. In physics this information is usually already available - systems being continuous or effectively discrete by inspection - but not in the study of psychological processes.) The associated pattern of phenomena can be modelled only poorly, and *ad hoc*, by combinatorics with a small finite set of states. With many states the fit improves, but not the motivation; a very-many-finite-state system can do so many things, so few of which match this. Only a continuous limit in the background can make the cusp reasonable.

Evidence for its occurrence, then, is evidence for the appropriateness of a continuous rather than a discrete picture of the process studied. (Though it does not exclude an *aspect* best treated as discrete, and for the brain's processing of words and pictures such aspects very likely exist.) There is evidence for its occurrence in nerve net dynamics, and some for its relevance to concept selection, such as the choice between the concepts 'man' and 'girl' above. Where should one look for evidence of its relation to language?

5. A 'bad' example

An outstanding current case of meaning reversal is the black American use of 'bad' to mean 'good'. The historical naturalness of this is obvious (when "The only good Injun is a dead Injun", the wise man admires those the oppressor does *not* call good; blacks have got wise), but not my concern here. The usage has become fashionable among white college students for their own reasons, and coexists with the earlier meaning. In some situations the two are morphologically distinguished (the superlative of one remains 'worst', that of the other is 'badest'), but in general there is a decision problem.

A touch of imitated black accent, or another word with black connotations shortly previous, or a tone of admiration, will disambiguate the word one way. Anger or a highly formal speech mode will disambiguate it the other. I suggest for experimental test that there is a grey region of tone values where the interpretation depends on context, just as there are neutral contexts in which it depends on tone; with both in the neutral range it remains ambiguous. Furthermore, the history-dependence of the interpretation is plausibly organised, like that of the man/girl figures above, by the cusp catastrophe. One may argue for this as follows.

I will represent a word not by a point but by a cloud of points, standing for the range of ways it can be spoken; for initial simplicity, a cloud in two dimensions, though far more than two variables are doubtless needed to specify a speech mode completely. (This is no real loss of generality, since the loop below could encircle the set of cusp points in any dimension as it does in two.) These clouds of points are mapped by the brain into a representation space of concepts, slightly different for different points in the same cloud. Some word-and-tone points can map to more than one concept point, depending on context, so this mapping (call it Φ) is one-to-many. For a particular word-and-tone, it gives the various concept points processing *can* lead to: they are not all realised every time. Identify two points in our 'input space' of words-and-tones if they have Φ-images, the resulting concept points in common or very close. Formally, an appropriate probabilistic theory or a development of the 'tolerance' notion would be helpful here; but there are problems with the brain papers of Zeeman (1961, 1965, 1968) using tolerance spaces, and my own development of the concept (Poston 1972) is not easily adaptable here. Informally, think of two word-and-tone points as identified if one could replace the other in a sentence with no perceived change of meaning, though the word in a transcription is different.

Drawing clouds as overlapping where thus identified, in white American college speech we have a ring as in Figure 9.

The vertical variable describes intensity of feeling, increasing downward, though the horizontal one is hard to label. 'Bad' overlaps at on end of its range with the unambiguous 'awful'; at the other, given positive tone or Afro-American context, with the equally unambiguous 'super'. It never coincides with mild or neutral words like 'so-so', but always expresses firm approval or disapproval. The middle range, around indifference, is inaccessible at the intensity level of 'bad'.

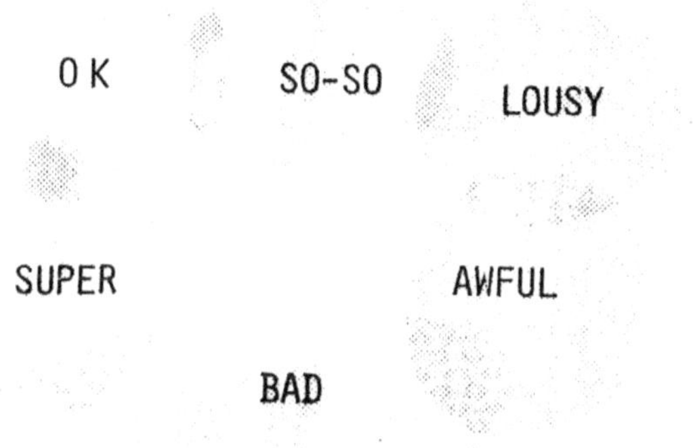

Fig. 9 A ring of words with overlapping meanings. Similar loops can be found in French ('terrible'-'super'-'chouette'-'pas mal'-'vache'-'epouvantable'-'terrible') and German (closing by the adverbial ambiguity of 'schrecklich').

Adding the approval/disapproval variable to the picture as a third dimension, the one-to-many character of the mapping is resolved (Figure 10).

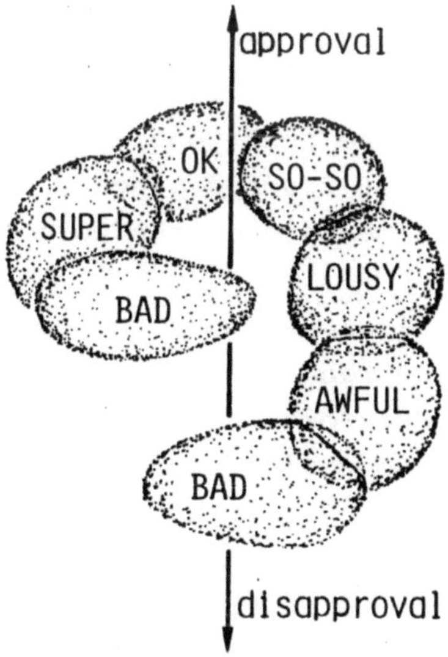

Fig. 10 The two senses of 'bad' separated by an additional variable.

If we follow reasoning like that in Poston and Stewart (1978b), with similar heuristics of a general nature, we are led to embed Figure 10 in Figure 11, where the point-clouds centre on the regions marked in the surface.

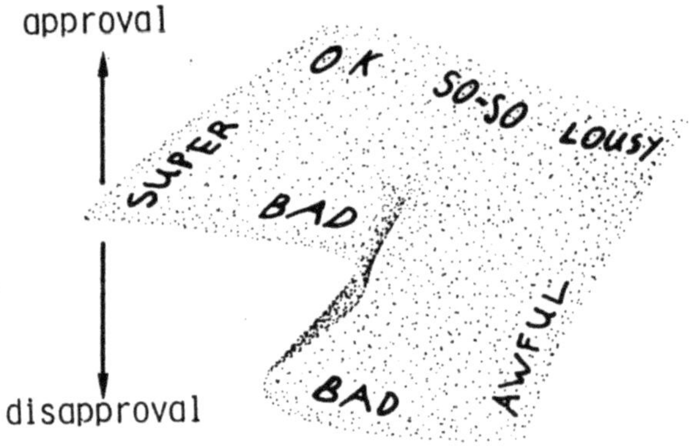

Fig. 11 A surface organising the concept clusters of Figure 10.

There is a continuum of conceptual states between the two components of 'bad', but since the middle sheet of the folded surface consists of unstable equilibria and not attractors, they are not available to this word. Dynamically, if we move 'bad' from an extreme negative tone to extreme positive we may expect a jump in perceived meaning as we meet the lower fold; returning, we do not jump back until reaching the upper one. There is a tone range in the middle where the concept reached depends on recent history. We have *hysteresis* in the meaning of the word.

Quantitatively, it is hard to be rigorous about the overlaps in the figure, for example between the ranges of 'so-so' and 'OK'. The rear parts of the graph may therefore seem little more than metaphor, but they have led us to something testable, in principle at least.

Restrict attention to 'bad' itself, as in Figure 12. The model does not merely predict hysteresis as we vary cues (tone or context) between positive and negative, with uncued, uncontexted 'bad' (e.g. written and in isolation) ambiguous by occupying the centre without a determining history.

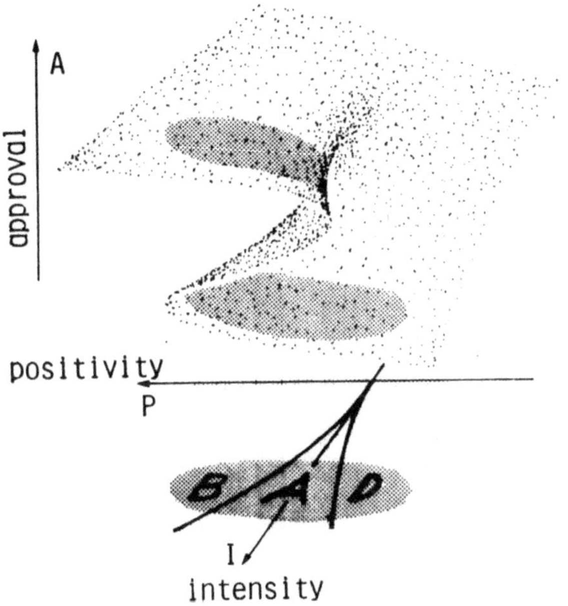

Fig. 12 The suggested graph of response to 'bad' with various levels of intensity and positivity cues.

It predicts that as vocal or contextual indicators for intensity of meaning become stronger, the region of ambiguity widens rapidly; if one is close enough to the organising cusp point, so that the characteristic local geometry dominates, the increase is faster than linear. (Near enough to the cusp point, asymptotically, bi-concavity is an invariant of cusp geometry, though it does not always hold globally. Hoping that asymptotics remain useful in the measurable domain is normal in the hard sciences and the very backbone of such success as the soft sciences' ubiquitous linear regression has. Globally, very few relations fit a linear model; asymptotically, most differentiable relations do. Near most points the surface in Figure 12 can locally be fitted fairly well to a linear dependence of A on I and P, around the fold and cusp points it cannot. Cusp-fitting is thus at least less local than plane-fitting, and exactly as general for similar reasons. The appropriate techniques for data-handling, generalising linear regression, are now being worked out; Cobb and Watson, 1980.)

The American college student being the second most frequent experimental animal in behavioural studies after the rat, experimental test of these predictions may be practicable. As regards tonal cues, such experiments would both require careful work on their parametrisation and be a sensitive test of technique in this neglected area, very likely cross-fertilising with the speech recognition problem. Even if the implications for language structure suggested in this paper are invalid, these cues are clearly an interesting and insufficiently studied set of signals important to day-to-day human communication. If, however, the non-linearity and analogue character of these signals is as strongly coupled to the structure of language as here argued, similar geometry in response to positivity and intensity cues in *written* material involving 'bad' should show up in experiments, if such cues can be suitably ranked. (Note that the prediction of a wider hysteresis/ambiguity gap for higher intensity is a non-parametric one, meaningful given only the unscaled rankings. This may make the design of suitable experiments more possible.) Such evidence for continuous geometry - even in this one example - would be of major importance as regards the structure of language, and the structures to be used in linguistics.

Confirmation of these predictions, or of similar phenomena elsewhere in language understanding, would not establish any particular continuous model of language, or validate the above chain of reasoning in any detail. The predictions gain force from the cusp's universality in non-linear continuous systems, but the same feature weakens it as a tool for discriminating between such systems. However, that chain of arguments chiefly led to a good-seeming place to look for characteristically analogue behaviour. If found, such behaviour has theoretical consequences, just as water quenches thirst whether found by dowsing or by scanning from orbit.

6. Conclusion

Many workers in 'Artificial Intelligence' insist that they are not trying to mimic the workings of the human brain, but to make problem-solving machines/programs. If a machine can win a game of chess, prove a theorem, or find its way around Mars or the New York subway, it need not do it the way a human would. (We have lots of humans to do it that way, already - though a robot goes cheaper to Mars.) The same applies here. If one can give either an abstract or a machine-implemented model of language - traditionally called a grammar - that successfully deals with disambiguation, pronoun referent identification and so forth, it does not matter if it does not describe brain workings too. Machine calculation with numbers does well *not* to imitate the process of a human doing arithmetic.

But the results of arithmetic are, in some touch-Gödel sense, objective. Human language is a human activity, and the right answers to questions of word meaning are whatever the community of human users says they are. ("The question is," said Humpty-Dumpty, "which is going to be the master - that's all.") Humans, it is argued above, assign meanings on an analogue basis, and have an irresistible tendency to give continuous aspects to meaning and sentence structure. They feel strangers and afraid in computer languages, with their resistance to nuance, metaphor and allusion, and given the chance would turn them into something different. American Sign Language, invented as a discrete set of symbols, rapidly acquired so many analogue, kinesthetic elements ('*far* that way') that transcription became a major problem, and adequate translation into American English as hard as for Chinese. A metaphor works when the concepts involved 'feel the same' to the visual/aural/kinesthetic imagination: is this combinatorial in character? (Consider "she gave me a severe case of whiplash of the heart." (Niven 1970), where the scene pivots on the fact that - to normal humans - the metaphor is self-explanatory.) To date, computer approaches to language analysis have made exactly no progress with metaphor.

The heart of metaphor being recognition of similarity, those workers in pattern recognition who seek to find a 'grammar' of images, modelled on few-finite-state grammars of language, may be importing a fundamental error about language into a field where it will be even less fruitful.

The task of a theorist, or machine analyst, of language is to predict the results of the human tumbling of superficially discrete words in a system of continuous dynamics. Partial successes with combinatorial approaches have been achieved, and more doubtless will be. But how far can an approach that negates part of the essential character of a process, not recognising the kind of thing it is approximating, succeed in describing its external behaviour? Is the finite state combinatorial approach to language bad, or bad?

Acknowledgements

I am grateful to John Kozak, Thomas Ballmer, Ian Stewart and Otto Rössler for their advice on this paper, and to Hermann Haken for the hospitality of his Institute and stimulating conversations.

Bibliography

Abraham, R.H./Shaw, C.D. 1982, *Dynamics: the Geometry of Behaviour*. Aerial Press: Santa Cruz

Bateson, Gregory. 1966, "Problems in Cetacean and Other Mammalian Communication". In: K.S.Norris (ed.), *Whales, Dolphins and Porpoises*. University of California Press: Berkeley, pp. 569-799. And in: G. Bateson, *Steps to an Ecology of Mind*. Paladin: Geneva, 1972, pp. 334-348.

Briggs, W./Newell, A.C./Sarie, T. 1981, "The Mechanism by Which Many Partial Difference Equations Destabilize". In: H.Haken (ed.), *Chaos and Order in Nature*. Springer: Berlin, in press.

Cowan, J.D./Ermentrout, G.B. 1978, "Some aspects of the behaviour of neural nets". In: I.A. Levin (ed.), *MAA Studies in Math 15, Studies in Mathematical Biology I*. MAA 1978, pp. 67-117.

Cobb, L./Watson,W. 1980, "Statistical Catastrophe Theory: an Overview". In: *Mathematical Modelling*, 1, pp. 311-317.

Fisher, G.H. 1967, "Preparation of Ambiguous Materials". In: *Perc. Psychophysics* 2, pp. 421-422.

Harth, E./Lewis, N.S./Csermely, T.J. 1975, "The escape of *Tritonia*: dynamics of a neuromuscular control mechanism". In: *J. Theor. Biol.* 55, pp. 201-228.

Levinson, S.F./Libermann, M.Y. 1981, "Speech Recognition by Computer". In: *Scientific American*, 244(4), pp. 56-68.

Niven, L. 1970, *Ringworld*. Ballantine Books: N.Y.

Magnus, R.J./Poston, T. 1979, "Infinite Dimensions and the Fold Catastrophe". In: W. Güttinger/H. Eikemeier (eds.), *Structural Stability in Physics*. Springer: Berlin, pp. 63-83.

Minsky, M./Papert, S. 1965, *Perceptrons*. MIT Press: Cambridge, Mass.

Poston, T. 1972, *Fuzzy Geometry*, Ph.D. thesis. Univ. Warwick.

Poston, T./Stewart, J.N. 1978a, *Catastrophe Theory and its Applications*. Pitman: London & San Francisco.

Poston, T./Stewart, J.N. 1978b, "Nonlinear Modelling of Multistable Perception". In: *Behavioral Sci.* 23, 5, pp. 318-334.

Zeeman, E.C. 1961, "The Topology of the Brain and Visual Perception". In: K. Fort (ed.), *Topology of 3-Manifolds*. Prentice-Hall: Englewood Cliffs, pp. 240-256.

Zeeman, E.C. 1965, "Topology of the Brain". In: *Mathematics and Computer Science in Biology and Medicine*. Medical Research Council, pp. 277-292.

Zeeman, E.C./Buneman, O.P. 1968, "Tolerance Spaces and the Brain". In: C.H. Waddington (ed.), *Towards a Theoretical Biology*. Edinburgh University Press: Edinburgh, pp. 140-151.

Zeeman, E.C. 1976, "Duffing's equation in brain modelling". In: *Bull. Inst. Math. Appl.*, 12, pp. 207-14. And in: E.C. Zeeman, *Catastrophe Theory: Selected Papers 1972-77*. Addison-Wesley: Reading, 1978, pp. 293-302.

CASE, AKTIONSART AND ERGATIVITY
A SEMANTICAL BASE FOR TYPOLOGICAL ISSUES

Thomas T Ballmer (+ 2.12.1984)
Sprachwissenschaftliches Institut
Ruhr Universität Bochum

The problem of finding the best method to establish a base for typological issues of natural language phenomena still remains to be solved. Two of the predominant methods are the *language extensive* method and the *theory intensive* method.[1] The first relies upon a stepwise investigation of some selected linguistic features through many languages. It is inductive. The second method relies upon theoretical insights. It is deductive. Both of these approaches can be quite effectfully criticised, the first as being ad-hoc with respect to the selection of the empirically investigated features since it is hopelessly overloaded with data, the second as being empirically fastidious and often only vaguely related to linguistic reality. The question is wether there is a third way of avoiding some of these criticisms while maintaining the better aspects of the two approaches mentioned here.

It is proposed in this article that such a "third way" exists,

---

1 The first says: select a certain linguistic feature and check through as many languages as possible whether or not it applies, and, after having done this for a great number of features, investigate how the results (for such different features) relate. Universal (generalized) properties are then conceived as implications generally true for all or most languages investigated. Typological properties are clustering features, characterizing subsets of languages within a system of such subsets. The second method says: try to understand what the nature of language is, i.e. what kind of entity language is, what requirements it should fulfil, what function it has, what purposes it serves. Universal properties are then to be taken as deducible from the general truths of the theory thus obtained. Typological properties are not thus derivable, but differentiate different (non-arbitrary?) sets of languages.

which is more than a simple additive mixture of the two known methods. It lays the base for many issues of universal grammar and typology. This method starts out from a large scale semantical investigation of linguistic material and thereby lays a fundament for linguistic theorizing: *semantical space*. Knowing the structure of such a *semantic space* of natural language renders it possible to investigate grammatical matters of natural language on three levels: context and performance (pragmatics), content (semantics) and form (syntax, morphology). In this article we are particularly interested in typological matters cutting across these three levels of pragmatics, semantics and morpho-syntax and thereby dealing with central grammatical categories: *case, aktionsart*[2] and *ergativity*. In this connection, the notion of *case* occupies a dominant place both for typological issues as well as for revealing the tight relationship between context, content and form. *Case* thus seems to be a crucial matter for linguistics. This is the more so if taking into account how successfully the notion of case has been able to resist an adequate definition or even a theoretical assessment. As it turns out the above-mentioned semantic space - i.e. the structured domain of what can be semantically expressed in and by language - relates the linguistic forms of case-markings to the underlying relevant contextual processes, i.e. mediates morpho-syntax and pragmatics in a categorially and hence typologically relevant way.

In order to accomplish the task set forth in this article, the following problems have to be settled: 1. How do we arrive in a principled and empirically justifiable way at disclosing the structure of the semantic space of natural language? This would direct us towards an alternative approach to provide a base for typological issues. 2. What is the present state of case definition, is it satisfactory or not, if not: what would be a better proposal? Providing a successful alternative for a case definition would allow us to get a hold on one of the most central and yet

---

2 We use the German term 'aktionsart' (corresponding roughly to process phase) in order to avoid confusion with English or Russian aspect.

very poorly understood grammatical categories relevant for typological issues. As we shall be able to see, an answer also leads us further along our path via the question: 3. What are the consequences which result from the "third way", the method advanced in this paper, for typology, universal grammar and the theory of grammar?

In the sequel we are planning to deal with these three major groups of problems.

1. The Semantic Space of Natural Language

Very briefly we shall try to *motivate* why we should set out to seek the semantic space of natural language. Firstly, is it at all possible to do so? It would not be, of course, if we had no reason to assume that there were such a space. Secondly, is it at all feasible? It would certainly not be were we not to provide an appropriate method of discovery. Thirdly, is it useful for our purposes? If not, we should, then, rather refrain from engaging ourselves in the troubles of such a task.

1.1 Properties of the Semantic Space

Before answering these three questions we should first get a rough idea of what the properties of such a semantic space of language would be. As we have briefly mentioned, a semantic space is a structured domain of what can and should be expressible in and by language. It is, in other words, an organized domain of the anthropologically relevant meanings. The existence of such a domain provides us thus with the limitations of expressibility for human language. Assuming, as I think correctly, that there may be radical differences of expressive power among the various languages, the semantic space should, in addition, characterize the possibilities of such differences and, specifically, organize the natural growth or change of the expressive power of a language through history.

This means that the semantic space, if it exists at all, must be a rather deep sitting *universal*, true for all languages, throughout their historical development. It would not come as a surprise then, that essential aspects of the semantic space are *genetically co-determined.*

The centrality of semantic space is documented also by another fact. Being the universal matrix for semantic expressive power, the semantic space should also have its impact upon morpho-syntactic expressive power: what linguistic forms do exist in what number, how arbitrary and/or principled are their specific meanings? Only the meanings demarcated by the semantic space are prone to being communicable and thus conventionally related to certain linguistic expressions. In other words, the semantic space severly restricts, and thereby determines, the morpho-syntactic forms. But the situation is even more refined and subtle. Roughly speaking, only those forms relating to meanings which are relatively easily localizable in semantic space are liable to exist stably. Accordingly, the semantic space even marks out which meanings are more likely to be expressed linguistically and which are not. Because there are anthropologically more relevant meanings than others, it does not come as a surprise that the differences in lexicalization, length, formal and phonetico-phonologic simplicity, frequency of occurrences in a text, etc. are related to the intrinsic shape of the semantic space. The more relevant a meaning, the more succinct the expression (this is a brief version of what we call the *principle of relevance).*

That the anthropological relevance, i.e. a contextual or pragmatic property, has a bearing upon which and how meanings of the semantic space are expressed indicates a crucial fact: the semantic space mediates between what is the case contextually (pragmatics) and what are the linguistic means (morpho-syntax). The principle of relevance tells, in an overall and general manner, how the relation between pragmatics and morpho-syntax is stabilized within semantics.[3]

As things are, a major shift in linguistic theorizing is perceptible. For hard core linguists, especially structuralists and generative grammarians, morpho-syntax lies at the centre of interest and of linguistic theory. For applied and hyphen linguists, pragmatics plays that central rôle of orientation. With reference to the universal nature of *semantic space* and its primordial relation, with respect to syntax as well as pragmatics, a different stand has to be taken. *Semantics* seems to lie at the very core and centre of language and hence of any adequate form of linguistics. If this were a tenable position it should be a relatively easy matter to adduce further and independent arguments for substantiation.

In fact there are *four* novel and independent arguments supporting the centrality and significance of semantics for linguistic analysis. *First*, the communicative purpose argument: speakers subjectively assume that the purpose of language and linguistic communication is to transport *meanings* (and not, say, to exchange meaningless codes). Psycholinguistic memory and operation tests (paraphrasing, abstracting, expanding) attest these findings on a quite objective level. *Secondly*, the translation argument: what matters essentially for an adequate translation is usually the content, i.e. something semantical. Matters of style, rhyme and expression are, except for special situations, of a rather marginal relevance. *Thirdly*, the language learning argument: whether in

---

3 The *principle of relevance*, in one of its more elaborated forms, reads as follows: In its use, human language adapts itself to the anthropological needs of its users. The more relevant something is anthropologically, i.e. for the users of the language, the more overtly, i.e. briefly, succinctly, conventionalized, abundantly etc., it is expressed by a linguistic entity (-type). The basic level of overtness, i.e. brevity, succinctness, convention, numerousness etc. is *lexicalized simple non-technical words*. Comment words thus constitute the fundamental level of linguistic expressions as regards anthropological relevance. (This could be taken as a relevance-linguistic definition for linguistic level of words.)

first or second language learning, it is of primary importance to learn how to express one's intentions. To be able to express what one means, temporally precedes and is of higher priority than to express this in a morpho-syntactically and stylistically correct way. This is true for the three year old as well as for the foreigner. *Fourthly*, the invariance argument: meanings are (quasi-)invariants. They do not much vary through speakers (interpersonal quasi-invariants), through languages (linguistic quasi-invariants) nor through time (historical quasi-invariants).

This being so, our motivation to seek to prove the existence of a semantical, hard core structure, as the *semantic space* represents it, is reasonably substantiated. If semantic space exists and if it could be found, it will certainly prove to be useful: as the core structure for a demonstrably central topic with respect to syntax and pragmatics, namely for semantics.

1.2 Retrieving the Semantic Space

The task is now to find out the form and structure of the semantic space. There are two major problems to be dealt with. Firstly, this semantic space is a *holistic* entity. It relates to language as a whole. It is not possible to pick out an arbitrary part (or a fragment) of language and to hope that a part of the semantic space will reveal itself. Secondly, the semantic space is a *universal* entity. It thus relates to language tout court and not to a particular language. Accordingly, the question arises of how one can possibly find out something reliable, non-hazardous and representative about that huge and invariant underlying object.

The following assumptions are now made:

(1.1) In order to retrieve the semantic space, we consider the expressive means of languages in their (proto-)typical readings and apply a universally invariant method of semantic classification.

(1.2) Because the lexicon provides the linguistic (i.e. semantic, syntactic and pragmatic) basis for the remaining expressive means of language, we concentrate upon the lexicon, and extend our knowledge about the semantic space later on from that basis.

(1.3) As a first approximation, it suffices to concentrate upon the expressive means (i.e. the lexicon) of one or a few modern cultural languages (like English, Chinese, Arabic, Russian, French, German). The argument in support of this relies upon two facts: semantics and even more so its core, the semantic space, is quasi-invariant, as we have been stating. It may vary slightly, but not considerably, through languages. Secondly, there exists a translation culture between all linguistically investigated languages equalizing the meanings expressable in different languages, i.e. rendering them comparable and, thus again, quasi-invariant (cf. that the same thing could by no means have been said for syntactical or properly pragmatic phenomena).

Thus we break down our investigation of the holistic and universal semantic space to the prototype semantic analysis of linguistic expressions, more precisely the lexicon, of one single language, without any loss as we have been arguing. Once we have found the form and structure of the semantic space, we shall be in the position to reconstitute, more or less easily, the modifications necessary for adapting larger sets of expressive means and other languages . The latter claim rests upon the two arguments of the (quasi-)invariance of semantics and of the intertranslatability of linguistic expressions.

We may ask, then, whether it is possible to further reducing the expenditure without essential loss of structural information about the semantic space. This question can in fact be answered in the positive. Not the entire lexicon of a language is needed for the analysis in question, a carefully selected fraction will do. The question is which one. A list of arguments will now be put forward in support of the claim that it is sufficient to concentrate upon the *verbs* of a modern, cultural language.

(2.1) The verbs are grammatically central. As predicates, they are traditionally seen as the major meaning carriers of the sentence. Nouns, adverbs, adjectives are traditionally seen to be less relevant for the meaning determination of sentences. This fact is reflected in the rôle the verb plays in approaches such as traditional logic and predicate calculus.

(2.2) Verbs play a central rôle in the syntax of sentences. This is documented by a number of facts. Verbs determine the syntagmatic relations in a sentence, the case or actant frame and important types of subordination. This rôle is reflected in the structure of theoretical approaches such as in valency grammar and dependency grammar.

(2.3) Moreover: verbs are numerous enough to reveal the shape of the semantic space, but not too numerous to forbid an empirical study. Verbs are, on average, less complex in meaning than nouns, which are usually polyfunctional. Verbs are monofunctionally designating processes (including actions, speech actions) or dispositions for processes. Nouns, on the contrary, designate forms, functions, purposes, (sometimes) processes, concrete and abstract objects of various sorts. Adjectives and adverbs are less fundamental than verbs and nouns. In addition, verbs are usually not very ambiguous. Verbs are historically stable in meaning as well as in number.

(2.4) Verbs are thus semantically and syntactically central and exhibit a high degree of semantical stability. They also exhibit a property more important than many others valuable for revealing the semantic space: they are *semantically representative*. This cannot be proven here. Only a more thorough and systematic lexico-semantic analysis of the entire thesaurus of a modern cultural language, together with a theoretical underpinning of the empirical findings, could demonstrate that point.

Thus we are in the position now to say that the verbs of, say, German, more precisely, the German non-composite standard verbs[4] constitute the rather large empirical basis[5] for finding out the structure of the semantic space. The method of performing this task has been described in a number of places.[6] Thus we shall concentrate upon the essentials when explaining some of the methods for arriving at the structure of the *semantic space* via the verbs.

1.3 The Structure of the Semantic Space

We have been arguing that we should start out from the verbs in order to arrive at the structure of the semantic space. We shall order and arrange them according to a method which is itself independent of the language chosen and hence universal. Its results may, however, depend upon the language chosen. Accordingly, the set of all verbs can be ordered by (essentially) only two semantic relations in the realm of competence of the native language user: meaning similarity and meaning presupposition. The set of all verbs (about 21,000 German verbs, about 8000 simple standard verbforms, about 13,000 simple standard verb meanings) can be categorized by the meaning similarity relation into about 40 models. The categories in a

---

4 i.e. more specifically the non-prefixed verbs (except those prefixed with *er-*, *be-* and *ver-*) and non-composite verbs (like *dreh-bohren*) and the non-technical, non-dialectical, non-idiosyncratic verbs. All of those do not contribute to the *basis* semantic space. Their meaning is derivative from the more elementary verbs.

5 There are about 8 000 such German non-composite standard verbforms, and about 13 000 such verb meanings.

6 cf. e.g. Ballmer/Brennenstuhl 1978; Ballmer/Brennenstuhl 1981a,b; Ballmer 1983

model can be ordered by the presupposition relation, (nearly always) linearly, and the models among themselves likewise linearly. The result is the following three-dimensional verb-thesaurus structure:

(3)

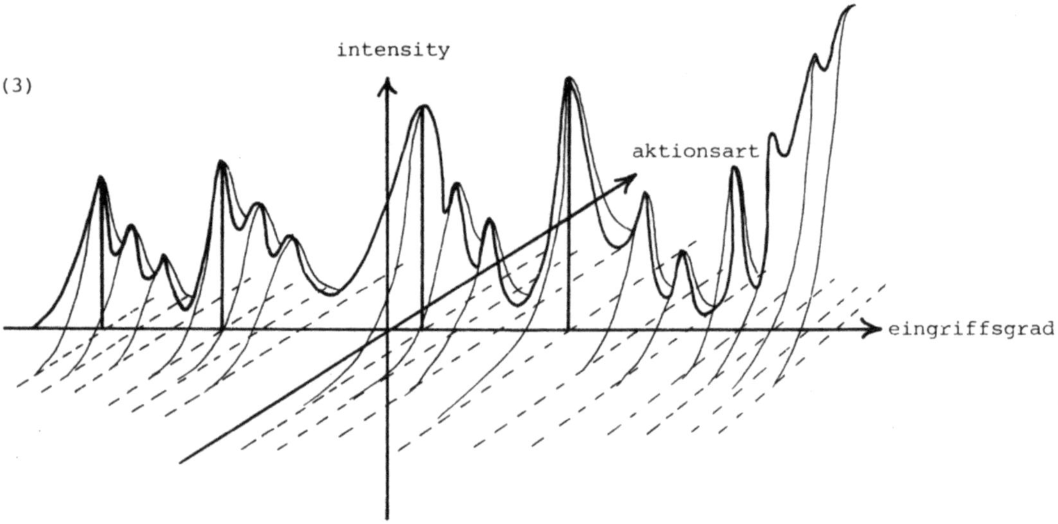

Fig. 1

The remarkable fact about this structure is its simplicity and its regularity (cf. Ballmer/Brennenstuhl 1981), especially its three-dimensionality. The three axes are the eingriffsgrad (degree of influence), the aktionsart (not to be confused with aspect) and the activity intensity. The eingriffsgrad gets its name from the distance and efficiency range (reichweite) of the control the subject has.[7] The first axis, the *eingriffsgrad,* is a measure of (fugitive or resistant) survival potential against attacks upon the object/organism in question.[8] The

7 The eingriffsgrad dimension is morphologically represented in Arabic. The verb root inflection shifts the meaning of the resulting expression along the eingriffsgrad dimension (e.g. causativation etc.)

8 The presupposition order of the verb categories makes it overt that the entities involved as subjects of the process designated by a certain verb in question get more and more control over their environment along the eingriffsgrad dimension.

eingriffsgrad correlates with the process of bio-physical evolution (cf. Ballmer 1981b) and neuro-physiological morphology (cf. Ballmer 1981a). The second axis, the *aktionsart*, is a measure of the development of a process (beginning, ongoing, ending; or in linguistic terms: inchoative, durative, terminative). The aktionsart correlates with the (onto-)genesis of objects, organisms, actions, etc.[9] Along the aktionsart dimension are situated the closed processes with a natural beginning and end, whereas along the eingriffsgrad dimension, there is the one prominent open process: cosmological bio-genesis, with the typically human activities at the top end. The third axis, the *activity intensity*, is a measure of the subject (of the process designated by the verb). At the beginning of the process the activity is low, in the middle there is a maximum, towards the end it falls again. The resulting activity curve resembles a hat for which reason it gives rise to what is suggestively called hat structure. As an example, consider the following few hats:

(4.1)

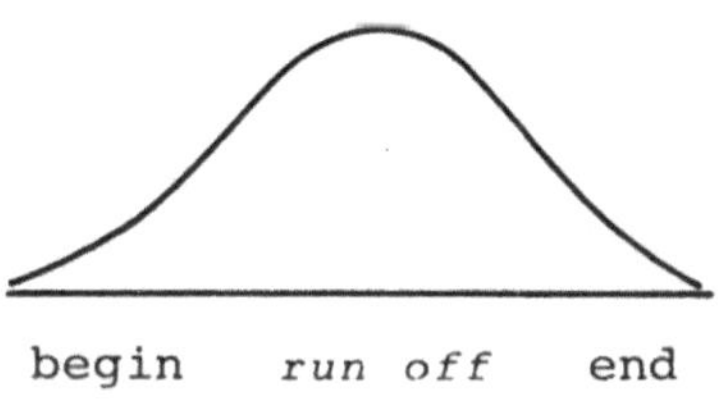

(4.2)

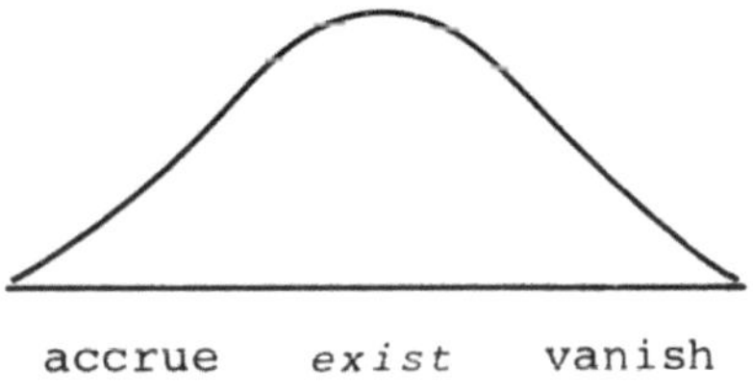

(4.3)

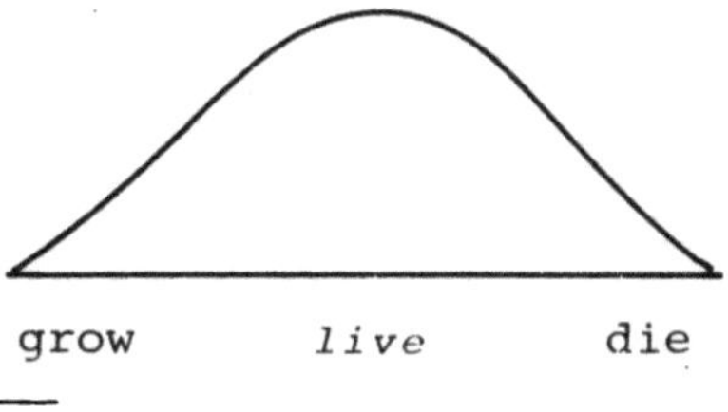

---

9 The aktionsart dimension (inchoative, durative, terminative) is morphologically realized in Russian.

(4.4)

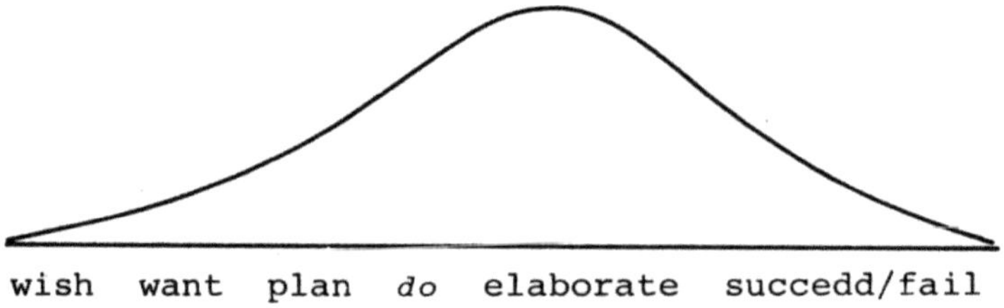
wish want plan *do* elaborate succedd/fail

(4.5)

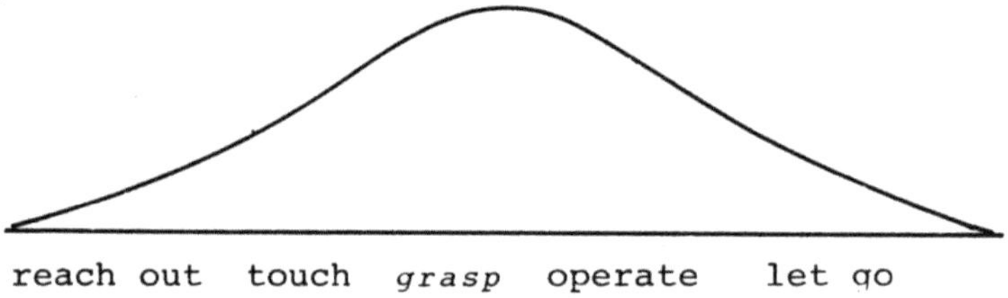
reach out touch *grasp* operate let go

(4.6)

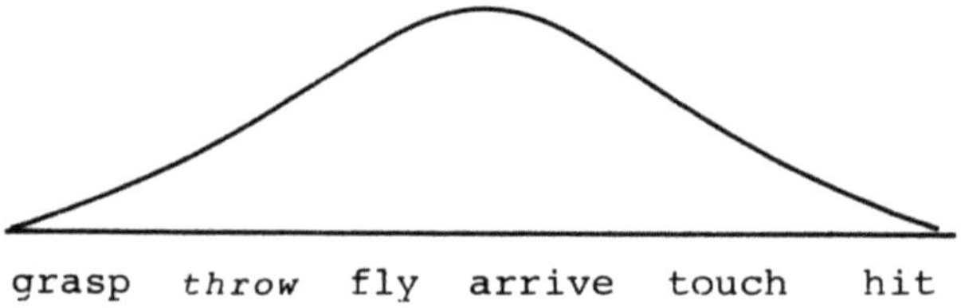
grasp *throw* fly arrive touch hit

(4.7)

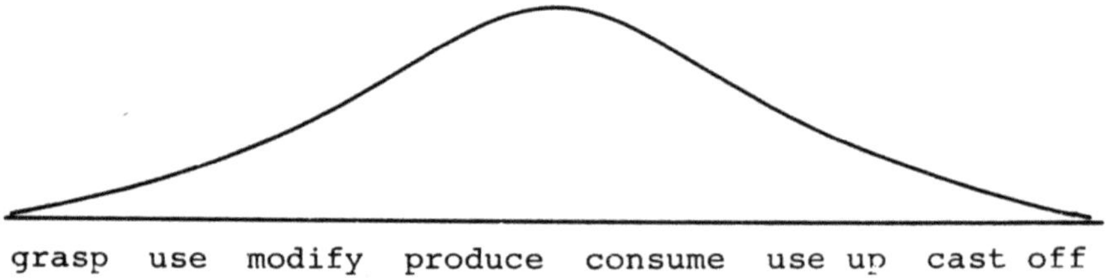
grasp use modify produce consume use up cast off

(4.8)

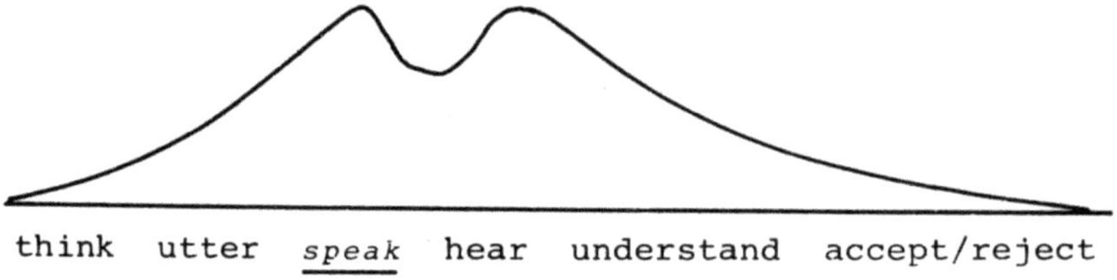
think utter *speak* hear understand accept/reject

Here follows a list of the forty-two [normal] plus four [speech act] models to give an idea of the socio-culturally relevant processes (of the German-speaking person):

(5) *German Verb Models*

SV	State of Affairs Model
VO	Process Model
VP	Model of Special Processes
EX	Individuals-Objects-Existence Model
ER	Properties and Relations Model
VH	Behavior Model
LE	Life Model
LG	Life Model for Social Groups
BW	Motion Model
ZS	Experience Model
PW	Passive Perception Model
PK	Psychological Effects Model
BE	Effecting Model
HD	Action Model
AB	Active Movement Model
FB	Locomotion Model
AW	Active Perception Model
IF	Information Model
BT	Work Model
DR	Execution Model
ÄU	Utterance Model
PS	Psychological Activities Model
WA	Danger-Fear-Risk Model
PZ	Process Control Model
FO	Reproduction Model
GR	Grasp Model
FH	Guide Model
US	Support Model
FR	Freedom Model
TR	Transport Model
FD	Drive Model

TZ	Compose/Decompose (Manipulation) Model
ES	Substitution Model
BA	Modification Model
PR	Production Model
KO	Consume Model
NG	Give and Take Model
KA	Buy and Sell Model
RE	Regeneration Model
TA	Transaction Model
BH	Housing Model
BS	Visit Model
SP	Speechact Modelgroups:

EM	Expressives (Emotives)	IA	Interaction
EN	Enactions (Appeal)	DC	Discourse

In order to focus upon the essential form of the classification result, we now list the more prominent models in the appropriate hierarchy (the shell model, cf. Ballmer/Brennenstuhl 1981b).
The hierarchy of prominence is numerically justified. The prominent models are those containing, relatively speaking, a maximum number of verbs. The exciting fact is that numerosity (of the verbs in a model) correlates with the relevance we are tending to attribute to the process semantically expressed by the model (principle of relevance, cf. Ballmer/Brennenstuhl 1981b).

(6) The Shell Model of the Anthropologically Relevant Processes

1.	first major shell:		being	(nonactive)
1.1	first shell:	VO		(processes)
1.2	second shell:	ER		(objects)
1.2.1	first subshell:	EX		(inanimate)
1.2.2	second subshell:	LE		(animate)

2.	second major shell:	influence	(active)
2.1	first shell:	HD	(action)
2.1.1	HD (action)		
2.1.2	FO (dislocation)		
2.1.3	BT (play and work)		
2.2	second shell:	GR	(grasp)
2.2.1	GR (grasp)		
2.2.2	TR (transport)		
2.2.3	TZ (manipulation)		
2.3	third shell:	BA	(modification)
2.3.1	BA (modification)		
2.3.2	NG (give and take)		
2.3.3	TA (marketing)		
2.3.4	(or 2.4?) Speechacts		

1.4 Processual Bifurcation

This thesaurus structure now exhibits a series of further remarkable properties. One of the linguistically preponderant facts is the intransitive - mesotransitive - transitive - hypertransitive distinction parametrized by the verb-thesaurus structure:

(7)

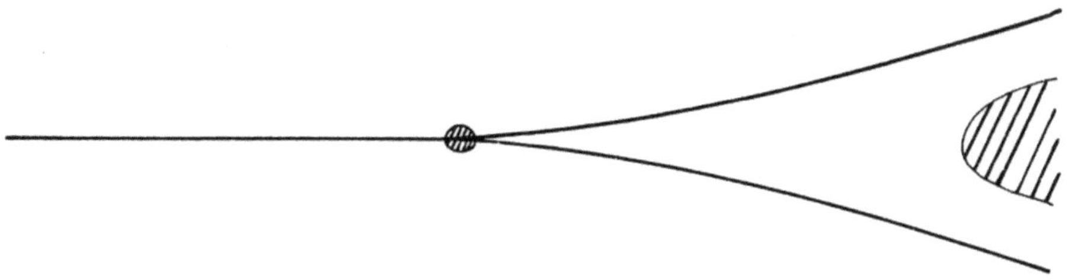

VO EX LE HD	FO GR	TR BA	SPA
intransitive	*mesotransitive*	*transitive*	*hypertransitive*
(e.g. act)	(e.g. grasp)	(e.g. throw)	(e.g. speak)

The lower models, namely the process model (VO), the existence model (EX), the life model (LE), the action model (HD), the locomotion model (FB) and the activity model (BT), all contain verbs which are (or can be) only intransitive (*begin, run off /elapse, proceed/, exist, live, die, want to, act, run, go, arrive. sleep, play, work)*. The running off of the process takes place at exactly the point where the subject is. The subject wanders along with the processes he controls (or which happen to him) through the aktionsarten (inchoative, durative, terminative):

(8)

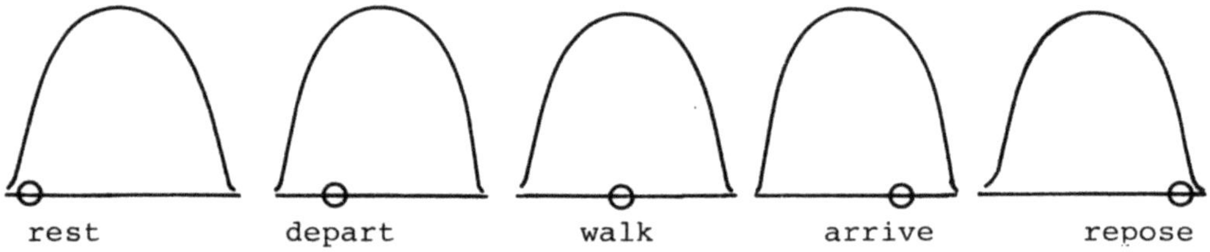

This changes at the hybris point (reproduction model) and, briefly afterwards, at the grasping model (GR, *reach out, touch, grasp, operate, let go, draw back*). It becomes especially clear in the case where *operate* is specialized to *throwing:*

(9) *Throwing* (Sample sentence: *Danny throws the ball)*

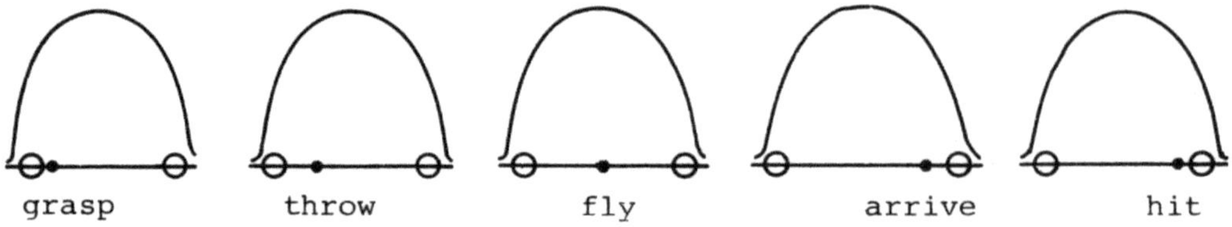

We observe that the subject is only initially involved in the process he causes: *grasp, throw*. The ongoing process *(fly, arrive)* only concerns the object thrown *(ball)* and in the last phase (*hit*) a target is involved.

The models above the hybris point (the point of minimal transitivity!) are transitive. In contrast to an intransitive process, a transitive process leaves its originator (the subject) and takes place at a different entity, the object, and may even affect a further entity (the target, goal). Transitive processes go over (transive) from the source (subject) to the goal (indirect object) mediated by a vehicle (object).

Hypertransitive processes have a still larger range. The mediating process (like e.g. *fly* in the case of the transitive throwing) has no name. Compare *speak* or *send (a letter)* for which there is (in German) no conventionally used verb for the travelling of the sounds or the letter, respectively.

We see here that the physical, biological and even the social dynamics underlying the language, (and more specifically the verbs thereof) exhibit a clearcut structure. It is the structure of an immense, dynamic process starting out with initial cosmogenesis (VO-model of processes) leading to the formation of objects (EX-model of existence), and of living beings (LE-model of life), continuing with action (HD-models of action), grasping (GR-models of grasping), transport (TR-models of transport), modification (BA-models) and the linguistic models (SPA-models: emotion, enaction, interaction and discourse).

The prominent, dynamical structure of the semantic space, spanned by the verbs, is thus the fork branching at the hybris point. The fork characterizes the gradual process of subject/object separation beginning at the hybris point and proceeding to the grasping, modification and creation of new objects (including language and organisms, offspring).

(10)

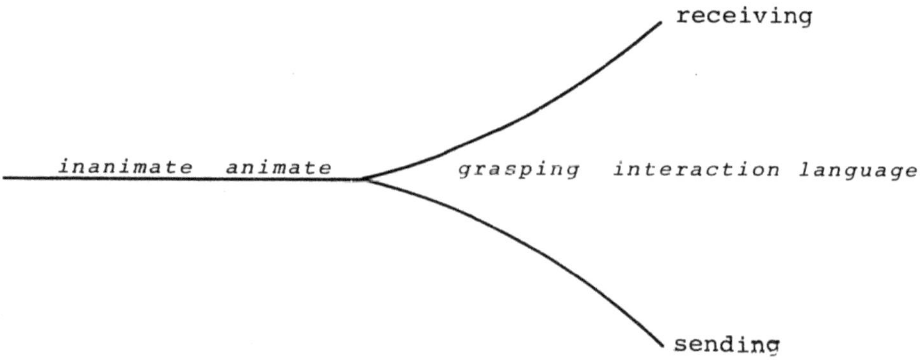

We call this therefore the *fork archetype*. In physics, and in Thom's catastrophe theory, this elementary archetype is well-known: it is the archetype going along with hysteresis, or, in other words, with the cusp catastrophe (cf. Thom 1971, Wildgen 1979, and cf. also Ballmer 1981a, Ballmer 1981b) or some higher catastrophe (e.g. the butterfly).

(11)

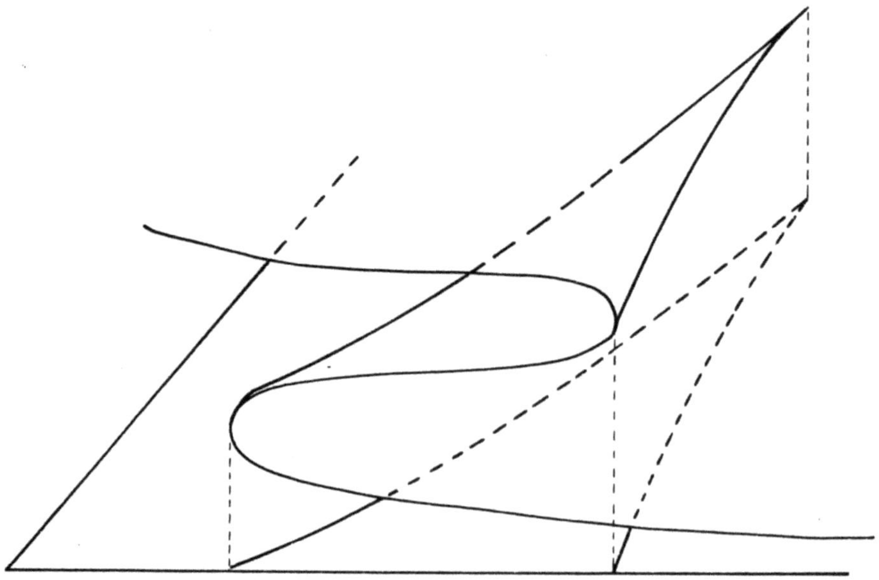

Fig. 2

2. Case Theory

2.1 About the Inadequacy of some Case Definitions

Traditional case definitions are of two kinds. They are either morpho-syntactical, and exclude any reference to semantical notions and hence lack of any content, or, they are semantical, in the sense that they refer to so called case rôles (or simply: rôles), and then lack any straightforward relation to the expressive means of the language in question.

Thus Duden (1973), e.g., gives the following definition:

(12) The *case* and *rôle* of the noun in the sentence is thus characterized: as subject it stands in the nominative, as object in the accusative, dative, genetive etc.

As it turns out in the subsequent text, cases are conceived exclusively, syntactically and on a purely formal level in accordance with what is governed by the verb. No semantic characterization of the notion of case is given and, in particular, none that would differentiate between the various cases. Specific case "definitions" turn out to be nothing but circular. The notions of case and rôle are used indiscriminately without any obvious reason.

Dirven/Radden (1977) propose the following definition:

(13) A case or semantic rôle is a complement determined by certain features standing in a constant relation to the predicator.

It does not seem to be clear what 'constant relation' means: constant with respect to which variation? The positive aspect of this definition is certainly its explicit relation to semantics. Being in the vein of Fillmore (1968), Dirven/Radden's approach, however, exhibits similar deficiencies. Thus it is fatal for both approaches not to carefully distinguish between cases and rôles. We observe again an indiscriminate use of these notions. These are quasi-synonymously used and confused as is demonstrated by the following argument.

Dirven/Radden define the dative thus:

(13.1) The rôle dative designates the receiver or beneficiary of an action designated by the predicator, or the person who experiences or owns something.

And Fillmore:

(13.2) Dative is the case of animate being affected by the state or action identified by the verb.

However it is easily shown that the same case is related to a considerably large number of rôles. A dative rôle, hence, cannot exist, and a dative is certainly not a rôle:

(14.1)	Der Bräutigam schenkt seiner Braut einen Ehering. (The bridegroom gives to-his bride a ring.)	*Benefactive*
(14.2)	Die Einbrecher raubten dem Rentner seine letzten Ersparnisse. (The burglars stole to-the pensioner his last savings.)	*Malefactive*
(14.3)	Dieses Kind ähnelt seinem Großvater (This child resembles to-his grandfather)	*Object of Comparison*
(14.4)	Mir ist kalt (To-me is cold, I feel cold.)	*Experiencer*
(14.5)	Erna traut ihrem Partner nicht. (Erna does not trust to-her partner.)	*Reference-Person*

(14.6) Du bist mir einer. *Point-of-View*
(You are for-me a-special-person.)

(14.7) Du kannst mir mal ... *Lokativ*
(You can at-me do certain things.)

(14.8) Der Bademeister droht den Kindern. *Affected Object*
(The bath attendant threatens to-the children.)

Dirven's definition of the dative rôle is fallacious, because it cannot said to be an exhaustive definition. Fillmore's definition is not in accordance with the facts. Both claims follow from the examples given in (14). In view of the examples (14.8, 8', 8'')

(14.8') Der Stein fällt runter *Affected Object*
(The stone drops down.)

(14.8'') Klaus traf den Lehrer *Affected Object*
(Klaus hit the teacher.)

the same rôle may even be expressed by different cases (dative, nominative, accusative). This leads to the statement that cases and rôles cut completely across. They have nothing to do with each other: the same case may relate to considerably different rôles, the same rôle to considerably different cases. A way out of this dilemma is wanting.

2.2 Processual Case Theory

2.2.1 Cases as a Relation between Aktionsarten

Having thus seen the inadequacy of the more traditional case conceptions, we are invited to come up with something more valid. For the reasons given, we cannot restrict our attention to the morpho-syntactical level (circularity reproach), nor should we confuse cases with rôles (inadequacy reproach). Having argued at length for the existence of the semantic space, we should suppose that a universally prominent relation, such as the *case relation* should immediately be connected with semantic space. This will in fact turn out to be precisely so.

The surfacial reflex of case is, as has been elaborated more or less correctly by the tradition, a relation between the verb and the noun, more precisely between the verb- and the nounphrase. It should not come as a surprise, then, that the semantic fundament for this syntactic reflex is a relation between what a verbphrase denotes and what the nounphrase in question denotes in a semantic space. The problem to be solved here, is, however: what do verbphrases and nounphrases denote with respect to the semantic space we were able to find? Verbphrases (verbs), we have seen, denote process phases, i.e. aktionsarten, with respect to a finite and rather limited set of verbmodels. Thus *beginnen (begin)*, *ablaufen (run off)* and *enden (terminate)* designate typical phases of the *Process Model* (VO), *ausstrecken (reach out)*, *berühren (touch)*, *fassen (grasp)*, *verwenden (employ)*, *loslassen (let loose)* designate typical phases of the *Grasp Model* (GR).

Nounphrases denote objects. In the process related, i.e. the processual conception we are advocating here, we should be more explicit. However, nounphrases denote objects when being processually relevant. *Jonas slings a stone at the window* implies, that with respect to the process of slinging, the stone is immediately affected and thus relevant, whereas the window is less so at that instant. *The stone* is in the accusative case. Only after the slinging has taken place and the stone has been slung and has flown through the air is it the case that *the stone hits the window*. Then the window is directly relevant. Only with respect to the process of *hitting* is it the window which is the object being immediately affected and thus directly relevant. *The window* is then in the accusative case. The adequate semantics for nounphrases is, hence, not static but rather, dynamic. A nounphrase is not assigned an object (or set of objects) invariably, but rather in a processually dependent manner. The processual phase matters within which the object(s) denoted become/s/ relevant. The relation between the central processual phase, indicated by a sentence /and denoted explicitly by the verb of the (simple) sentence/ and the phase in which the object(s) denoted by the nounphrase become/s/ relevant, is a very characteristic feature. It grasps the temporal relation between what happens as a process and the relevance phases of the object(s) involved in that process. As it turns out, there are three major *cases* in this relation: *preceding, simultaneity* and *postceding.* The processual phase in which a noun-denoted object may become relevant can be *before, simultaneously* or *after* the central process phase denoted by the verb of the sentence. These are three distinct cases of processual relation. These are, in fact, *the* three (oblique) *cases* we are looking for: *genetive, accusative* and *dative.* We should aim to solve the following problem on this basis. Given a simple sentence consisting of a verb(-phrase) and some noun-phrases, what case markings do the nounphrases get with respect to their meaning, the meaning of the verb and an underlying processual analysis based upon our knowledge of the semantic space? An adequate case definition should provide a solution to this problem.

Thus we define, less informally:

(15) *Processual Case Definition*

0. A linguistic case is primordially a semantic relation. This semantic relation is expressed morpho-syntactically in many languages, if not in all, by inflections (case markings, attached to the noun, more rarely to the verb), by prepositions or by syntactic constructions.

1. A linguistic, oblique case is a relation between process phases (aktionsarten) denoted by the two major linguistic phrase categories: verb(phrases) and noun(phrases).

2. With respect to a simple sentence S consisting of a verb

V and various nounphrases NP_i, a case is a relation between the process phase φ_V, denoted by the verb of the sentence and the process phase φ_{NP_i} in which the objects denoted by the nounphrase in question become relevant.

3. The condition for the three oblique cases is as follows:

$$\text{genetive} \equiv \varphi_{NP_i} \text{ before } \varphi_V$$

$$\text{accusative} \equiv \varphi_{NP_i} \text{ simultaneously to } \varphi_V$$

$$\text{dative} \equiv \varphi_{NP_i} \text{ after } \varphi_V$$

4. The typical, surfacial reflex corresponding to the three conditions under 3. are the three oblique case markings: genetive, accusative, dative. Moreover, there are further options of linguistic expression available when taking into account prepositions, or even verbs or syntactic constructions.

2.2.2 The Morpho-Syntactic Impact of Processual Case Theory

What is exciting about this definition is that it allows one to assign case markings once the processual situation and its thematized parts, by means of a verb and corresponding noun, have been given. Consider, *first*, some verb-controlled *genetive* constructions.

(16.1) Lehrer Böck erfreut sich des Weines.
(Teacher Böck enjoys himself of-the wine. (gen.))

(16.2) Der Enkel erinnert sich seiner Großmutter.
(The grandchild remembers himself of-his grandmother. (gen.))

(16.3) Der Priester gedenkt des Verstorbenen.

(The priest makes mention of-the deceased. (gen.))

(16.4) Tell harrt des Gessler.
(Wilhelm Tell awaits of-Gessler. (gen.))

In all cases, the object designated by the noun has been relevant [in a processual phase] before the phase denoted by the verb. In (16.1) the wine is presupposed to have been available before Böck enjoys it, in (16.2) *remembers* presupposes a phase of existence of the grandchild's grandmother, in (16.3) mention is made of the deceased whose existence is presupposed to have been relevant earlier. Even in a case like (16.4) where it is hoped that Gessler will appear and will thus to become relevant for Tell in the future (cf. aufwarten/auflauern jd1 jd3, wait somebody to-somebody (dat.)), Gessler has been relevant to Tell before. Thus it is highly interesting to compare the four cases (16.4) and

(16.4') Der Boss erwartet seine Sekretärin jeden Moment.
(The boss awaits his secretary. (acc.))

(16.4'') Der Boss wartet seiner Sekretärin auf.
(The boss waits/attends to-his secretary. (dat.))

(16.4''') Der Boss wartet auf seine Sekretärin.
(The boss waits for-his (dat.) [secretary (acc.)])

The closeness, in the three latter cases, between the expecting and the realization point is increasingly weaker. All of these cases stand in semantical contrast to (16.4) where *harren* refers not to the future event of encounter, but to the past cause for an encounter intention.

Secondly, we are interested in verb-controlled *accusative* constructions.

(17.1) Reinhold besteigt den Himalaya.
(Reinhold climbs the Himalayas. (acc.))

(17.2) Die Polizei übergibt das Geld.
(The police hands over the money. (acc.))

(17.3) Der Junge traut sich nicht.
(The boy dares himself (acc.) not.)

The accusative case is in use independently of whether the process phase designated by the verb is at the end (17.1), in the middle (17.2) or at the beginning of an aktionsart hat, i.e. independently of a verb having a terminative (resultative), durative (transactional) or inchoative (e.g. optative) aktionsart. The only thing which matters is that the entity designated by the direct object noun (*Himalayas, Geld, sich*) is relevant in the same process phase as the corresponding verb.

Thirdly, we are interested in verb-controlled *dative* constructions. Simultaneously we are trying to investigate the mutual relationship between different ways of analysing the same processual situation.

(18.1) Erika gibt die Tücher ihrem Kind.
(Erika gives the cloths (acc.) to-her child. (dat.))

(18.2) Erika behängt ihr Kind voll von/mit Tücher(n).
(Erika drapes her child (acc.) full of/with cloths. (gen.))

(19.1) Der Arzt verschreibt dem Patienten Pillen.
(The doctor prescribes to-the patient (dat.) pills.(acc.))

(19.2) Der Arzt stopft den Patienten voll mit Pillen.
(The doctor stuffs the patient (acc.) full of/with pills. (gen.))

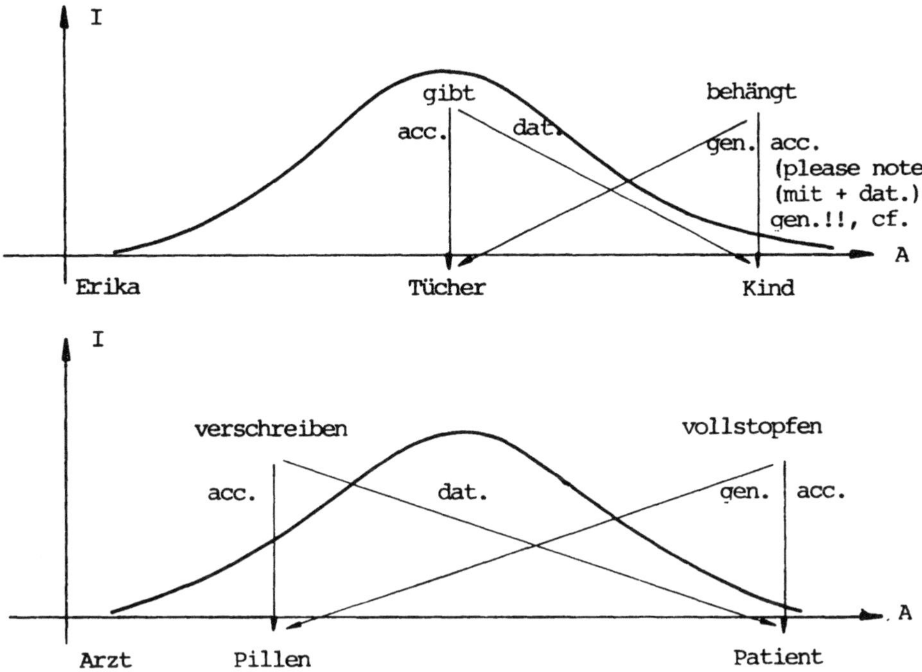

Fig. 3

The corresponding processual graphs demonstrate very clearly the relational character of cases in the way we have been describing it in (15) and more specifically in (15.3). These relations are again independent of the exact position of the relevant process phases with respect to aktionsart. *Dative* objects are relevant ahead and *genetive* objects behind of the verb-designated process phase. This is even true if the same process is analysed in two (or more) different ways: cases are *invariants* of a higher sort.

2.2.3 Case Refinements: Prepositions

This approach is rendered even more fruitful when taking into account alternative surface realizations of cases, namely those involving prepositions. According to the three major cases *genetive, accusative, dative*, there are three classes of prepositions, genetive, accusative and dative prepositions:

(20)

type (according to a rôle analysis)	genetive prepositions: proper	genetive prepositions: quasi-	accusative prepositions	dative prepositions
local	genetive von (by, of, from) aus (from)	instrumental	durch (through) via (via)	nach (to, towards); in (location / direction) (in); zu (to) auf (on)
in-strumental		mit (by,with)		
comitative		mit (with)		
authoritative		mit (with)		
objectual		mit (with)		
aiducative (helping)				für (for)

On this basis, entire classes of sentences are morpho-semantically relatable. *Von, aus* and genetive are in a sense equivalent (21), *via, durch* and accusative also (22), and so are *für, nach, zu, in* and dative (23), the difference being in the coupling strength between verb-designated process and noun-designated object (cf.20). The examples are ordered by increasing strength. The bare, i.e. non-prepositional cases represent the strongest coupling (cf. 23.1 through 23.5).

(21.1/2/3) Er kommt aus dem Büro./Er kommt aus dem Büro./
Er kommt des Wegs.
(He comes out-of the office./He comes from the office./
He comes of-the path.(gen.))

(22.1/2/3) Der Zug fährt via Chiasso/durch Chiasso./
Der Zug erreicht (fährt) Chiasso (an).
(The train goes via Chiasso/through Chiasso./
The train hits//arrives at/ Chiasso. (acc.))

(23.1) Er wirft den Ball für seinen Hund.
(He throws the ball for his dog.)

(23.2) Er wirft den Ball nach seiner Freundin.
(He throws the ball to his girl friend.)

(23.3) Er wirft den Ball zum Tor.
(He throws the ball to the gate.)

(23.4) Er wirft den Ball in den Teich.
(He throws the ball into the pond.)

(23.5) Er schickt den Ball seinem Hund. (Swiss German)
(He throws the Ball to-his dog. (dat.))

Thus also mixed cases of quasi-paraphrastic alternatives may arise, where some cases appear in the pure form, others in the prepositional form.

(24)

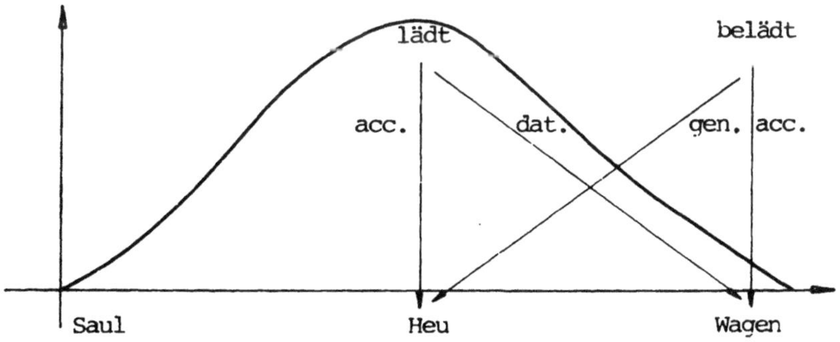

Fig. 4

(24.1) Saul lädt Heu auf den Wagen
(Saul loads hay on to the truck.)

(24.2) Saul belädt den Wagen mit Heu.
(Saul loads (effectively and completely) the truck with hay)

That the preposition *with* is a variant of the genetive case in the sense we have been defining here is easily demonstrated by pairs of quasi-synonymous sentence pairs.

(25.1) Der Kühler des Autos streifte die Straßenbahn.
(*The car's radiator* grazed the tramway.)

(25.2) Der Kühler vom Auto streifte die Straßenbahn.
(*The radiator of the car* grazed the tramway.)

(25.3) Das Auto streifte die Straßenbahn mit dem Kühler.
(*The car* grazed the tramway *with its radiator.)*

or, another set of quasi-synonymous expressions:

(26.1) Haarausfall kommt vom Alter.
(Loss-of-hair comes from aging.)

(26.2) Haarausfall kommt mit dem Alter.
(Loss-of-hair comes with aging.)

The prototypical meaning of *with* is not instrumental, but considerably more general. This is impressively illustrated by the sentence

(27) Saul belädt den Wagen mit Fritz mit der Gabel mit Heu.
(Saul loads (effectively) the truck with Fritz with the hayfork with hay.)

What matters is that "Fritz", die "Gabel", and "heu" are all relevant earlier than "belädt". *Aus, von* and proper genetive refer further back than mit (in German stemming from *mittels* meaning Mit-Teil-s, mittel-Teil-s, mittleres Wegstück, which is in turn related to causal weg-en).

2.2.4 The Casus Rectus: the Nominative Case

We should now ask whether it is possible to fit the nominative "case" into the framework set up here. Is it a case in the sense we have recognized that genetive, accusative and dative are, or is it, rather, not? In order to arrive at explicating a crucial notion of universal grammar, namely ergativity, we should know at our level and by our standard of explication what this is: the nominative case.

The nominative case is a case in the sense that it is a relation between process phases. It cannot, however be a relation between the verb-denoted process phase and a noun-denoted process phase. This is demonstrated by the following argument.

(28)

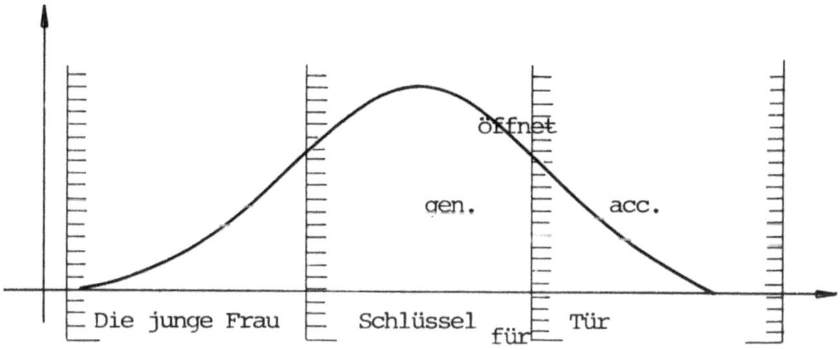

Fig. 5

(28.1) Die junge Frau öffnet die Tür mit dem Schlüssel.
(The young woman (nom.) opens the door (acc.) with the key (gen.))

(28.2) Der Schlüssel (nom.) öffnet die Tür.
(The key (nom.) opens the door (acc.))

(28.3) Die Tür öffnet (after the bell has been rung)
Die Tür öffnet sich.
(The door (nom.) opens /itself (acc.)/

In all three cases the process phase denoted by *öffnet* remains the same. Thus it plays no part in determining the marking of a noun as nominative. The relation between the *öffnet*-phase and the noun-denoted phase changes, but the nominative case relation remains the same. This proves that the nominative case cannot be of the same type as the oblique cases: it is not a relation between the two process phases in question.

The nominative is, rather, related to a topicalizing cut of the entire underlying process. A topicalization-cut is, thereby, a window starting at some point of a verb-model process and continuing to its end.

(29) *Processual Case Definition: Supplement for Nominative Case*

1. The nominative case is a relation between process phases.
2. The nominative case is a relation between the topicalization-cut phase and the noun-designated phase. As such it is independent of the verb-denoted phase.
3. The condition for the nominative case is as follows: If the noun in question designates the same phase as the topicalization-cut phase, the nominative case relation holds: nominative φ_{NP_i} simultaneously to $\varphi_{\text{top-cut}}$.

This definition holds for the cases displayed in (28) and many more. Observe that a noun-denoted processual phase may enter simultaneously to case relations. In this situation the corresponding noun has to be doubled: *Die Tür (nom.) öffnet sich (acc.).* (28.3).

2.2.5 Semantic Stability Requirements

Definition (29) has a number of further consequences. These may corroborate the decisions taken by it. *Firstly*, it is stated by the definition that case is a *relational* matter. Thus the very same pair of verb and noun (-phrase) may enter into different semantic relations depending upon the case relation holding between them:

(30.1) Otto tritt seiner Partnerin beim Tanzen auf den Fuß.
(Otto) steps to his girl-partner (dat.) while dancing on her foot.)

(30.2) Otto tritt seiner Partnerin beim Tanzen auf den Fuß.
(Otto steps his girl-partner (acc.) when dancing on her foot.)

(31.1) Nero opfert dem jungen Apollo.
(Nero sacrifices to the young Apollo (dat.))

(31.2) Nero opfert den jungen Apollo.
(Nero sacrifices the young Apollo (acc.))

The different case markings, expressing a different semantical relation, should hence lead to different meanings of the sentences. It has been observed by the traditional grammarians already that the dative is less obtrusive than the accusative (cf. Weisgerber 1958). Thus the person (Partnerin, Apollo) in these examples is less directly affected by the subject when using the dative rather than the accusative case. This corresponds also to the terminology indirect and direct object. (30.2) implies an intention by the subject, (30.1) does not, (31.2) implies the death of Apollo, (31.1) does not. According to the definition (29.3), cases are not defined as semantic rôles, nor with respect tu such. This also applies for (30) and (31). Only the relative simultaneity/ postcedence matters. In (31.2) the effect of the action described by the verb (death) is certainly much stronger than the effect described in (30.2). This is of no relevance to the case assignment, however. What matters is that both designate a relatively closer simultaneity than the respective dative cases.

Secondly, the processual case definition implies, that small variations of the meanings of verbs or nouns involved lead to comparable situations and, hence, equal case marking assignments. This is an expression of the continuity and semantic similarity of the underlying process and process phases.

(32.1) Olaf gibt seiner Mutter einen Apfel.
(Olaf gives to-his mother (dat.) an apple (acc.))

(32.2) Olaf schenkt seiner Mutter einen Hund.
(Olaf gives-as-present to-his (dat.) mother a dog (acc.))

(32.3) Olaf sendet seiner Mutter einen Brief.
(Olaf sends to-his mother (dat.) a letter (acc.))

(32.4) Olaf sagt seiner Mutter ein Wort des Dankes.
(Olaf says to-his mother (dat.) a word of thanks (acc.))

Conversely, the difference of case markings with respect to similar verbs (or nouns) should indicate a meaning difference.

(33.1) Mike hilft seiner Frau Ulrike
(Mike aids to-his wife (dat.) Ulrike.)

(33.2) Die Baföq unterstützt die Studenten.
(The Baföq subsidizes the students.)

(33.3) Madame aide Monsieur. (French)
(Mrs. helps Mr. (acc.))

(33.1) indicates personal help, which leaves with the helper some option of refusal. (33.2) is less personal and less cautious. It is official and therefore more effective. This accounts for the accusative. The French case (33.3) demonstrates that translation of cases requires an adequate translation of the verb: aider (French) seems to be closer in meaning to *unterstützen* than to *helfen*.

Similarly, there are meaning differences in (34) which are based upon the different underlying processual settings.

(34.1) Mir ist kalt.
(To-me (dat.) is cold.)

(34.2) Mich friert.
(Me (acc.) freezes.)

(34.3) Ich habe kalt. (Swiss German)
(I have cold.)

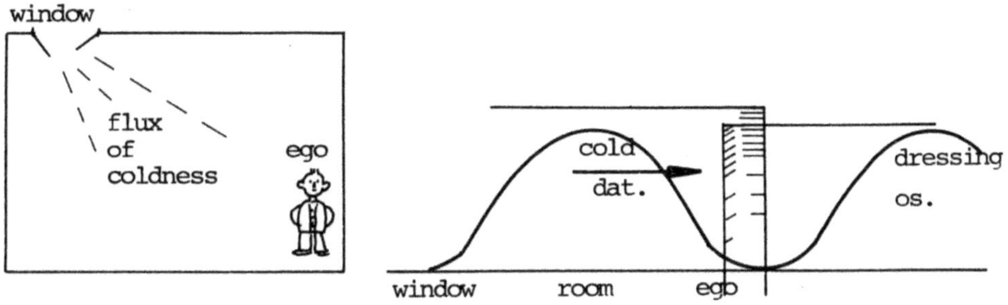

The real impact of a Processual Case Definition is that it relates processes going on in the world and the way such processes are quite closely articulated in language.

3. A Processual Analysis of Ergativity

Ergativity is characterized, as are many linguistic notions in two different ways, a morpho-syntactical and a semantico-pragmatic way. On the one hand there is thus the morphologic definition characterizing ergativity as identity of the case marking of the subject of the intransitive verb(s) and the object of the transitive verb(s), e.g. *mich ekelt (I am disgusted)* and *dieser Fisch ekelt mich (This fish sickens me)*. On the other hand there is the semantic definition that ergativity obtains if the subject is tied to a certain specific rôle, the patient (or experiencer). Bechert (1977) attempts to prepare a theory of grammatical categories which explains certain affinities among the grammatical categories and specifically ergativity. He attempts to do this on a (semantico-) pragmatical basis. He thus finds the following correlations.

(35.1) *Nominative System* (= Accusative System) - Animate NP - Subject of Transitive Verbs - Agent (as semantic rôle) - Definite/Generic NP - Topic - Dynamic Verbs - Present Tense - Imperfective Aspect - Singular - Active.

(35.2) *Ergative System* - Inanimate NP - Direct Object of Transitive Verbs - Patient (as semantic rôle) - Indefinite NP - Comment - Static Verb - Past Tense - Perfective Aspect - Plural - Passive

Heger (1982) demonstrates that his three actant functions, predicative function P, causal function C and finalist function F and their syntagmatic contrasts serve as a noematic, i.e. language independent tertium comparationis for accusative, ergative and active construction types. He thus corroborates the view that cases (language specific normal designations of predicative functions) are intimately related to an understanding of ergativity. He observes that the basic distinction between accusative system and ergative system consists in the fact that the following schema obtains (only the predicative and causal function are to be taken into account):

(36.1)

	Causal Function	Predicative Function
in syntagmatic isolation, corresponding to intransitive	C Nominative P	
in syntagmatic contrast, corresponding to transitive	C(P) Ergative	P (C) Accusative

As optimal solutions of the designative means the following three schemata are given:

(36.2)

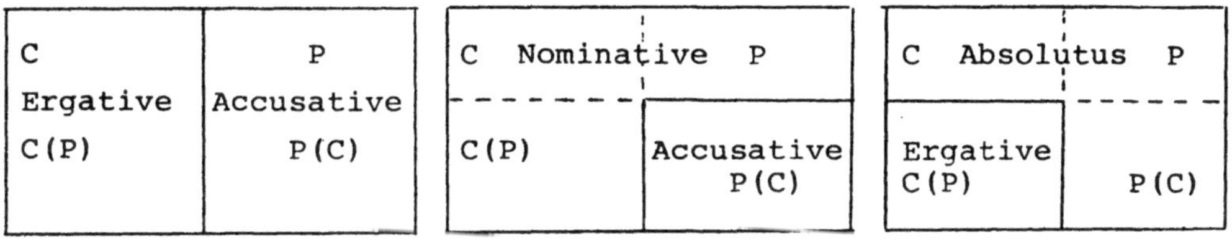

According to Klimov there is a complexity hierarchy of language types which parallels a historical relation: neutral languages, classifier languages, active languages, absolutive (ergative) languages, nominative (accusative) and alternatively to the latter dative languages. It seems that the languages of greater simplicity favor intransitive constructions, and that ergative languages are the first in this hierarchy to opt for transitive constructions.

The major question is now, whether there exists a perspective under which the results so far displayed are elucidated or even explained coherently. Is then a view from which the clustering of grammatical categories (Bechert), the noematic tertium comparationis (Heger) and the historical complexity hierarchy (Klimov), and possibly many other features of ergativity (e.g. that possessive constructions at the verb are often similar to ergative constructions, ergative languages rarely have a passive, but only an anti-passive

etc.) are seen to be the expression of one and the same underlying structure? The claim is that there is such a viewpoint, namely the perspective from the semantic space we have found when studying the verbstem meanings as a semantic system of natural language.

Let us begin the demonstration of this claim with Heger's noematic analysis. His schema (36.1) is in fact a binarily simplified version of the AxE plane of the verbthesaurus structure (semantic space):

(37.1)

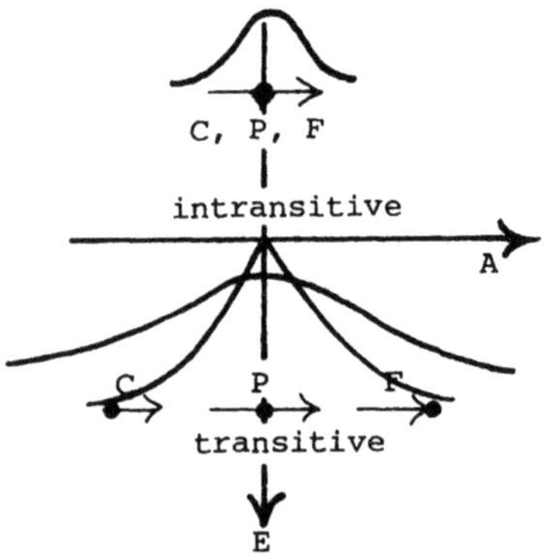

(37.2)

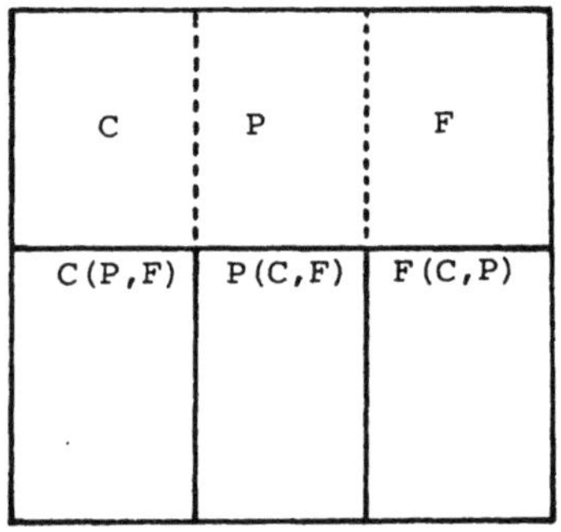

37.3)

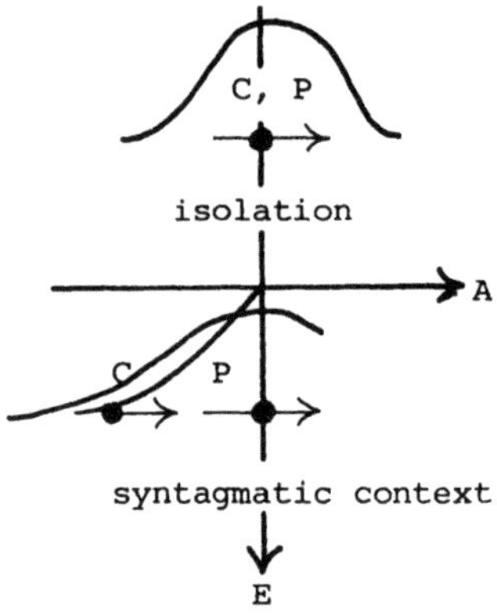

(37.4)

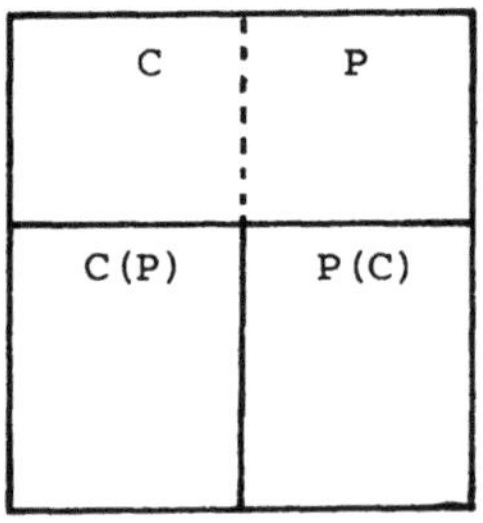

Klimov's hierarchy is also in immediate accordance with the semantic space. We may safely assume that the first processes getting lexicalized (by a verb) in a linguistic community are the most elementary ones, namely the intransitive processes, and only gradually, according to the principle of relevance, more complex processes are verbalized. Because of a natural persistence property of natural language, once it has begun to verbalize intransitive processes, it tends to verbalize also processes of transitive character (an agent is sending a transilient object to a goal or addressee) in an as intransitive manner as possible. Thus the central category of the verbmodel together with the then prominent thing, the object (Heger's predicative function), is getting verbalized and the "marginal" rôles are attached to the other bodies relevant before (Heger's causal function C) or after (Heger's finalist function F).[10]

(38)

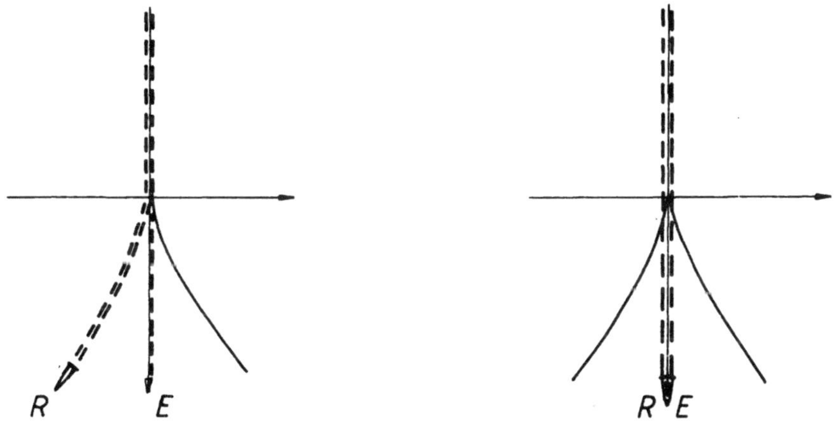

new rectality

nominative/accusative language

intransitive overgeneralization (old rectality)

ergative/absolutus language

---

10 The notion rectality designates the line of casus rectus.

The original rectality R along the semantically prominent central categories of the verbmodels breaks at one point down, because of its unstability caused by the complexity of the transitive model hats. The rectality could break towards the start (nominative/accusative languages) or the end/goal (dative languages) of the process hat. The latter case is definitely a quite rare case, for the end/goal of processes is often not too well distinguished. With reference to the model hat, the properties clustering together, that Bechert has observed, is often quite easily interpretable. In (39.2) the transitive situations are displayed, in (39.1) the corresponding intransitive situations.

(39.1)

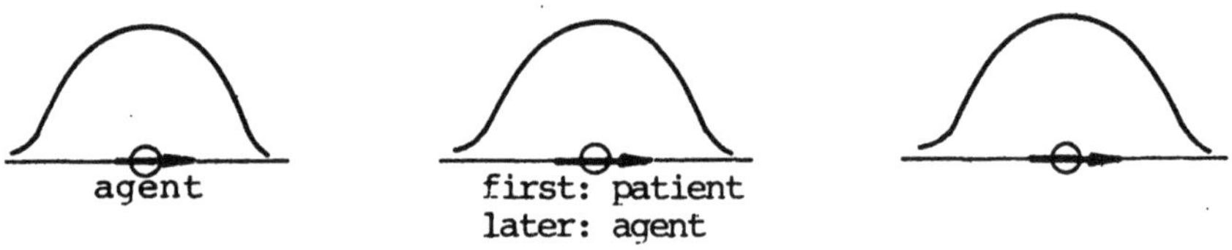

(39.2)

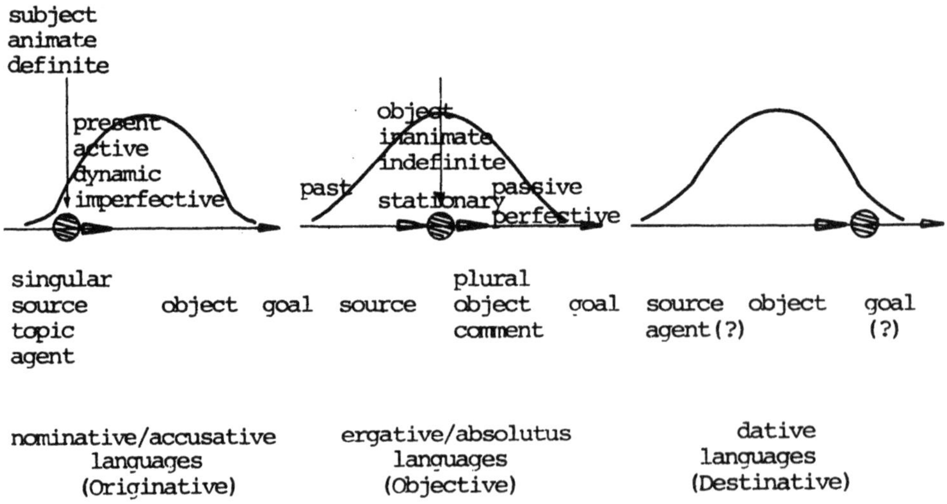

Thus it seems that ergativity is another grammatical category deeply entrenched in the processual reality underlying natural language. The cases discussed here can only indicate how a more thorough analysis could proceed and base our preliminary results upon a carefully elaborated philologic investigation.

(40)

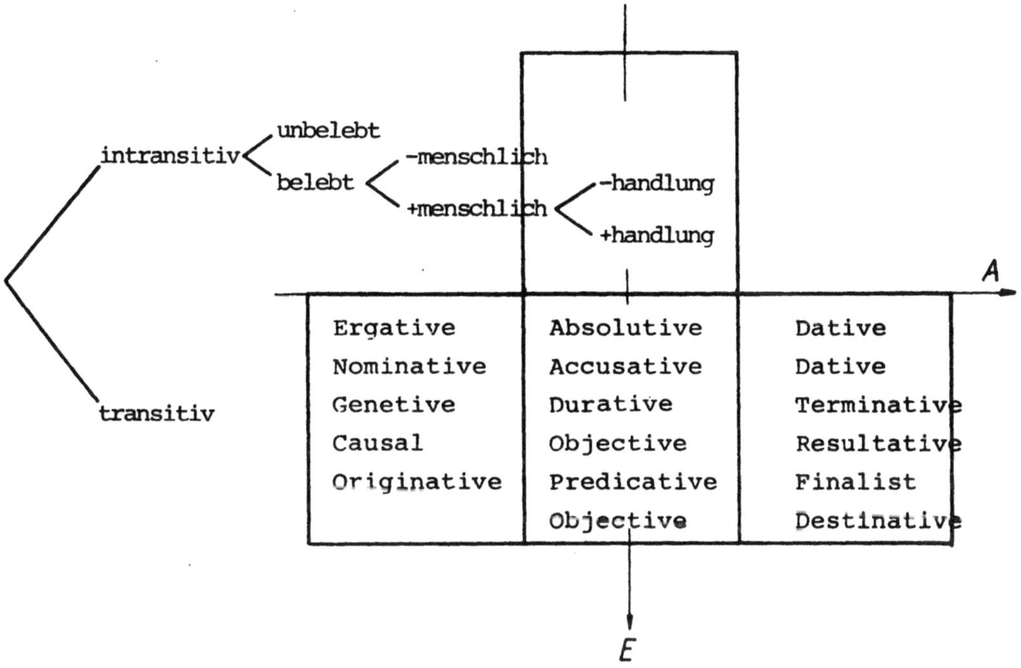

References

Anderson, J.M. 1971 *The Grammar of Case. Towards a Linguistic Theory*. Cambridge Univ. Press, Cambridge.

Ballmer, Th. 1981a "Neurobiologie und Lexikon". In: Schnelle, H. (ed.) *Sprache und Gehirn*. Suhrkamp, Frankfurt.

Ballmer, Th. 1981b "The Interaction between Ontogeny and Phyologeny: a Theoretical Reconstruction of Language and Mind". In: Koch, W.A. (ed.). *Semiogenesis*. Lang, Frankfurt, pp.481-544.

Ballmer, Th.; Brennenstuhl, W. 1978 "Zum Verbwortschatz der deutschen Sprache". In: *Linguistische Berichte*, 55, pp. 18-37.

Ballmer, Th.; Brennenstuhl, W. 1981a "An Empirical Approach to Frametheory: Verbthesaurus Organization". In: Eikmeyer,H.-J.; Rieser, H. (eds.) *Words, Worlds and Contexts*. De Gruyter, Berlin, pp. 297-319.

Ballmer, Th.; Brennenstuhl, W. 1981b "Lexical Analysis and Language Theory". In: Eikmeyer, H.-J.; Rieser, H. *Words, Worlds and Contexts*. De Gruyter, Berlin, pp. 414-461.

Ballmer, Th.; Dreher, Ch. 1983 "Der Semantische Raum, Linguistische Analysen zur Prozeßkombinatorik". In: *Conceptus*. Forthcoming.

Bechert, J. 1977 "Das Nominativ-Ergativ-Kontinuum und die pragmatische Fundierung grammatischer Kategorien". In: *L.A.U.T.* (Linguistic Agency, University of Trier), Series A, 47, Trier.

Dirven, R.; Radden, G. 1977 *Semantische Syntax des Englischen*. Athenaion, Wiesbaden.

Duden 1973, Grebe, P. (Hrsg.) *Duden, Grammatik der deutschen Gegenwartssprache*. Bibliographische Institut, Mannheim.

Fillmore, Ch. 1968 "The Case for Case". In: Bach, E.; Harms, R.T. (eds.). *Universals in Linguistic Theory*. Holt, Rinehart and Winston, New York.

Heger, K. 1983 "Akkusativische, Ergativische und Aktivische Bezeichnung von Aktantenfunktionen" in akup, 54.

Klimov, G.A. 1979 "On the Position of Ergativ Type in Typological Classification". In: Plank, F. (ed.) *Ergativity*. Academic Press, New York, pp. 327-332.

Plank, F. 1979 *Ergativity. Towards a Theory of Grammatical Relations*. Academic Press, New York.

Thom, R. 1975 *Structural Stability and Morphogenesis*. Benjamin/ Cunnings, Reading, Mass.

Weisgerber, L. 1958 *Verschiebungen in der sprachlichen Einschätzung von Menschen und Sachen*. Westdeutscher Verlag, Köln.

Wildgen, W. 1979 *Verständigungsdynamik: Bausteine für ein dynamisches Sprachmodell*. Habilitation Thesis, University of Regensburg. (The main chapters are published in: Wildgen,W. 1985. *Archetypensemantik*. Narr, Tübingen).

DYNAMIC AND ONTOLOGICAL FOUNDATIONS FOR A THEORY OF THE LEXICON

Wolfgang Wildgen
University of Bremen

The lexicon of a natural language, the lexical knowledge of a linguistic community and the lexical competence of individual speakers and hearers is normally considered as a finite set of entities (morphemes, words, idiomatic phrases). These entities can either be labelled as linguistic 'objects' or as procedures (a more advanced view proposed by Eikmeyer and Rieser, 1985). In either case it is not enough to list the elements, we must go further and specify the relations between these elements, their basic features, the levels of organization, etc. In seeking such an order of lexical items we may accept one of the following points of views and make them our guiding principle:

a) The internal structures are artefacts of the linguistic system and are conventional in the sense of Saussure's "arbitraire du signe" and the only general laws which obtain are laws of economy, hierarchy and mutual delimitation (cf. the field character of the lexicon). Candidates for organizational schemata are found in the domain of logics and basic set theory.

b) The internal structures of the lexicon are related to ontological structures, i.e. to:
   - the structure of the world, which determines parts of our lexical structures,
   - the structure of our sensory organs and processing mechanisms,
   - the structure of the socially constructed reality (especially if the labelled entities have no physical basis).

The two hypotheses are not independent as logical structures advocated in (a) are related to human cognition and are socio-historical constructs themselves. Thus (b) is the richer hypothesis. It is also the hypothesis we choose for our dynamic theory of the lexicon. The term "ontology" refers to this external correlates of lexical items and has at least three interdependent levels:

- physico-chemical ontology,
- sensory-cognitive ontology,
- social ontology.

The interdependence is bi-directional: on the one hand the physical world we label in our lexicon is mostly man-made and realizes plans on the level of our sensory-cognitive ontology and fulfils functions defined by the social ontology. On the other hand the social ontology has a material and economic basis, which garantees our physical existence and is perceived and categorized by individuals; the social ontology itself would not exist without individual psychic activities. This is not the place to elaborate on this complex interdependence; it should be born in mind, however, that *ontology* does not mean 'materialistic existence' as it covers all the levels mentioned above.

This ontology is not a "dead landscape" of entities, it is basically defined by processes which give rise to states, stabilize them and control changes and conflicts of states. The best candidates for general laws are the laws of motion (kinetics) and of morphogenesis (biological dynamics). In a sense our search for general laws is methodologically similar to the search for physical laws ever since Galilei. The term "dynamics" is therefore used in the same sense as it was introduced and developed by Newton, Leibniz and Kant (cf. Wildgen, 1985b).

For the general, especially the mathematical framework of the new dynamic paradigm I must refer to Wildgen (1982, 1985a) and Wildgen and Mottron (forthcoming 1987). In this article mathematical considerations shall be reduced to a minimum in order to describe the informal and intuitive content of the new approach and its application to the lexicon.

1. An informal sketch of catastrophe theoretic semantics

1.1 The general framework

We do not assume, as logical semantics do, that the universe is given as a system of individuals, attributes, and facts. The basic scope of semantics is represented by processes such as perception, recognition, and storage by the human organism; the recognition of qualitative change and of movements representing a central field. For the purpose of clarity, one may imagine semantics as an arsenal of abstract images which function either as invariants of perceived patterns or as invariants of self-evoked stimuli from the memory. Semantics, therefore, is no language substitute, as all semantic theories have in effect contended up to now; it is not language insofar as it distinguishes itself essentially from the level of realization, which is exposed to completely different restrictions by the efferent mechanisms. In a wider sense, one might also speak of iconic, or representational semantic theory.

We assume a stratification of *dynamic semantics*, which indirectly reflects different evolutionary strata.

(a) Semantic archetypes:
They comprise a small number of elementary propositional gestalts together with a hierarchy which allows inferences, metaphors and reductions in special contexts. The semantic archetypes contain minimal interpretations of biologically fundamental dynamic principles. We can call it the evolutionary "germ" of our language capacity. The minimal interpretation does *not* depend on specific cultural traditions, it constitutes rather a pragmatic universal of language and action. We shall give an illustrative fragment of archetypal semantics in the next

section.

(b) Semantic attributions:
The term "attribution" refers to "attribution theory" (cf. Heider 1958, Jones et al. 1971 and Harvey et al. 1976). Attributions are typical results of socio-psychological processes. The classification of interpersonal relations (Heider 1958), of motions, colours, the production of value judgements, stereotypes, attributions of motivation, responsibility and causation are typical examples. Together these structures constitute our supra-individual, social knowledge. The conventional and symbolic aspects of our languages have to be described at this level.

(c) Semantic elaboration and higher levels of text organization:
A certain degree of elaboration is conventionalized by our grammars and by the lexicon. We call it minimal elaboration. It may, however, also be controlled by the speaker according to the situation (cf. Wildgen, 1977b). If the elaboration does not correspond to the standards of the language (making sentences ungrammatical) we speak of reduction or ellipsis. In the second case the context implicitly contributes to the elaboration, in the first case the hearer must reconstruct the communicative intention of the speaker using his knowledge.

Our analyses refer exclusively to the level of semantic archetypes. The other levels require analytical tools that transcend those of catastrophe theory. They rather refer to psycho- and sociolinguistic theories.

1.2 Basic interpretations of dynamic configurations

Following the proposals of René Thom we assume two basic principles of interpretation which underlie our application of catastrophe theory to semantics:

(1) The attractors of the potential are interpreted as nominal entities: localities, qualities, phases or agents.

(2) The catastrophes (bifurcations and shifts of dominance) are interpeted as basic verbal structures.

In correspondence to principle (1) four types of interpretations can be distinguished (cf. Wildgen, 1982a, 25-29).

(a) The localistic interpretation. The attractors are interpreted as local areas (domains).

(b) The qualitative interpretation. The attractors are interpreted as domains in a quality space.

(c) The phase interpretation. The attractors are phases of basic movements or actions.

(d) The interaction interpretation (or agent-instrumental interpretation). The two attractors are interpreted as 'agents' ('participants' in Tesnière's terminology or 'semantic roles' in Fillmore's system).

Some examples taken from the interpretation of processes in the cusp-catastrophe will explain these notions.

In (a), (b), (c) we get processes of change between two symmetric domains, qualities or phases. In (d) elementary interactions between two agents can be derived.

<u>ad a</u>: The localistic interpretation

The slow dynamic along the path P can be related to a subject S, which travels on the path. We obtain:

(1) S moves from area A to B. We may, however, also concentrate on sub-areas of the process, thus obtaining:

(2) S leaves area A

(3) S enters area B

Fig. 1 shows the process along the path P (cf. Wildgen, 1982, pp. 42-46). The heavy lines are minima, maxima were omitted to simplify the graphical representation.

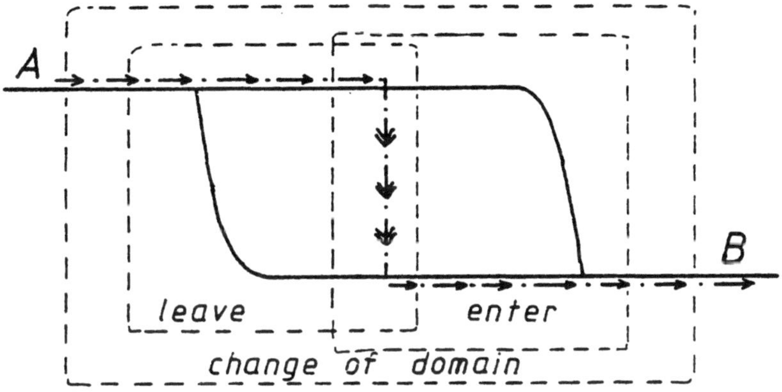

Fig. 1

ad b: The qualitative interpretation models bipolar quality oppositions which are very common in natural languages.

We shall illustrate the dynamics underlying such oppositions by describing the pair: asleep - awake. Figure 6 shows the process of awakening and falling asleep which is a "catastrophic" change in the bipolar dynamic field.

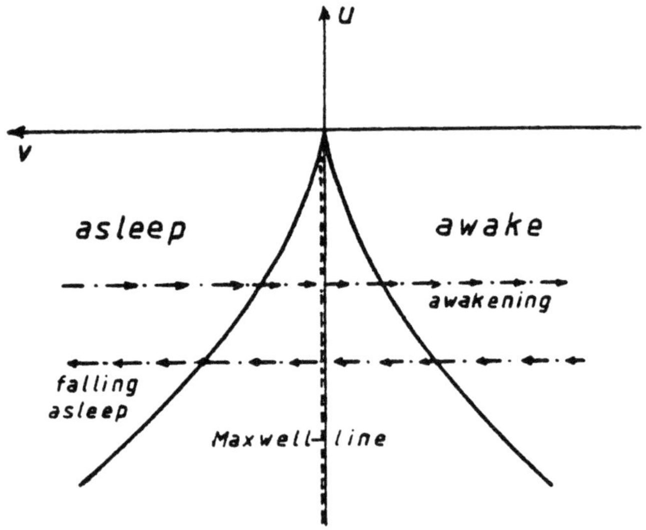

Fig. 2

ad c: The phase interpretation. We distinguish a neutral state of relaxation and a marked state of contraction (compare the concepts of "diastole" and "systole" in the dynamics of heart beating). The straight paths considered up to now are replaced by cyclical processes in the plane (x, v) of the cusp unfolding.

Rather simple realizations of this archetype can be found in sentences like:

S (subject) beats, S jumps, etc.

If the cycle is run through repeatedly, we get sentences like:

S (subject) waves, wags, etc. (iterative verbs).

ad d: The interaction interpretation. If we consider separately the left and the right side of Figure 1 we arrive at Figure 3 The corresponding archetypes are called: archetype of emission and archetype of capture.

emission
agent M_1 emits/throws/gives away/lets free/secrets agent M_2

capture
agent M_1 catches/takes/grasps/subjugates/absorbs/
devours agent M_2

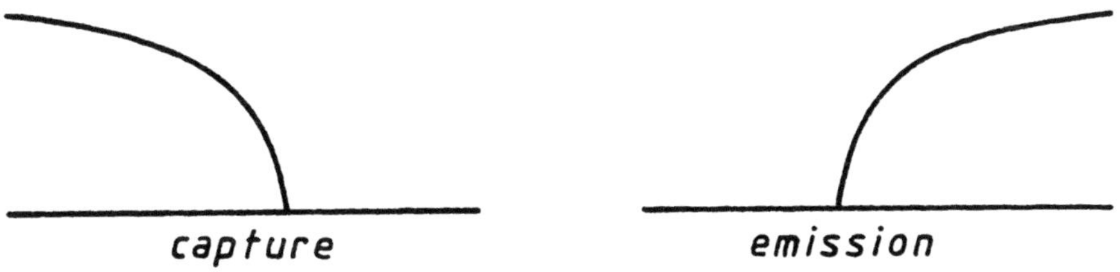

Fig. 3

The derivations sketched in this section were only meant to give an impression of catastrophe-theoretical semantics. The system is much more complex and has several levels. For fuller information the reader should consult Wildgen (1982 and 1985a).

1.3 The list of semantic archetypes

The elementary catastrophes (the fold, the cusp, the butterfly,...) used in the derivation of semantic archetypes can be hierarchically ordered by the relation of *bordering* (or 'abutment', cf. Gilmore, 1980, 135-140). This relation is obvious as it can be "seen" in the geometry of the unfolding. The "lower" catastrophes appear locally as singularities in the unfolding of higher catastrophes. Figure 4 shows the bordering diagram of simple unfoldings (in the sense of Arnol'd (1972)) up to codimension 4. The dotted lines relate this network to the nonsimple double cusp (X_9). For fixed values of the module K in X_9 we get also bordering relations. The double cusp is necessary to derive archetypes with 4 to 5 attractors (states, quality ranges, agents).

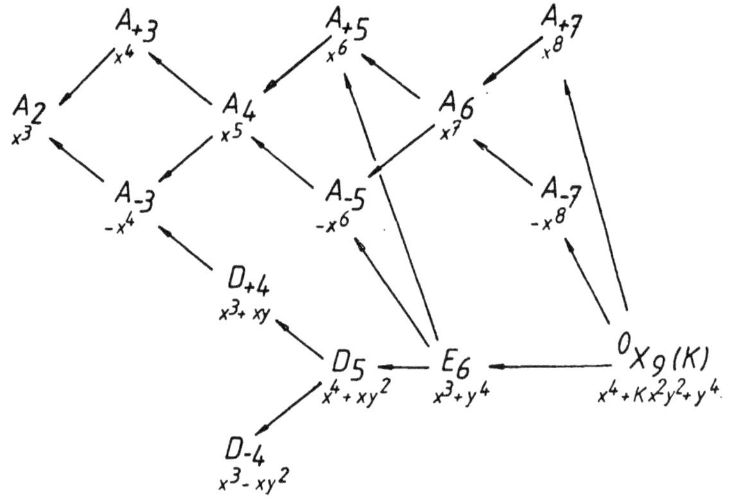

Fig. 4

Table 1 gives a short summary of the types of processes contained in the semantic archetypes derived at the different levels.

To the formal relations of bordering, which is a kind of topological inclusion, corresponds the relation of dynamic inference in C.T.-semantics. As the inference relates primarily to the dynamic pattern, we must control the filling of nominal roles and the elaboration of verbal dynamics to find concrete inferences between sentences realizing the semantic archetype.

The bordering relation is particularly simple if we compare those archetypes derived in the A_k-series. We shall, however, begin with an example of dynamic inference which connects the A_k- and the D_k-series.

name of the catastrophe	localistic	phase	possession/ control	interaction	instrumental
O-unfolding A_1	to be (in a domain)	to rest, to live	to have, to own	to be (in the scene)	---
fold A_2	to leave/ enter (a domain)	to die/to be born	to lose/ to find	to appear/ disappear (in the scene)	---
cusp A_3	to move (from A to B)	contract/ relax	to take/ to let have	to emit/ to catch	to beat/ to push
swallowtail A_4	semi-elementary archetypes involving a cascade, i.e. leave (renter-leave), change (return, change) etc.				
butterfly A_5	to travel (from A to B via C)	to pause (interme-diate state)	to give: A gives B to C	to propel (to a goal)	to hit, to touch (with an instrument)
elliptic umbilic (compactified in X_9) D_{-4}	to travel (A travels from B to C carried/escor-ted by D)	conditio-ned intermedi-ate state	messenger A sends B to C by D (=messenger)	secondary agent A affects B with C using D	secondary instrument A hits B with a gun D (propelling the bullet C)
D_{+4}, D_5, E_6 (compacti-fied in X_9)	further archetypes with 4 or 5 attractors. They are not treated in this report as their classification is still problematic.				

Table 1

(a) Dynamic inference from D_{-4} (compactified by X_9) to A_{+5}

From the compactified elliptic umbilic we derived the elementary archetype of the 'messenger'. From this archetype we can infer the elementary archetype of 'transfer'.

Examples

D_{-4}:	(a)	Tom sends	a letter	to Jim	by post	
		M_1	M_4	M_3	M_2	

A_{+5}: (b) Tom sends a letter to Jim

(c) Tom gives Jim a letter

Figure 5 shoes the dynamic relation between both processes. As the path in D_{-4} has to consider two internal variables, we had to simplify the diagram; the process along the path in the bifurcation set is symbolized by arrows.

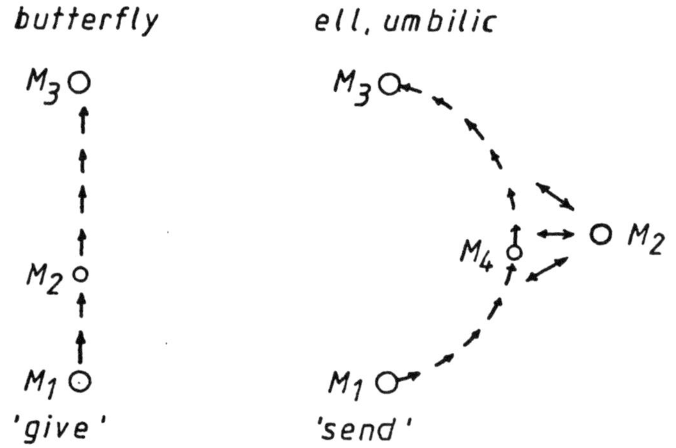

Fig. 5

We notice that the 'messenger' can be omitted; in (b) the choice of the verb "send" creates a background which implicitly refers to the archetype of 'messenger', whereas (c) does not contain this implicit elaboration. It seems that the lexicon of verbs has a kind of archetypal substructure which either enforces the choice of a correspondent set of agents or gives hints at presupposed but not realized dynamic features. We hope that a dynamic study of the verbal lexicon will be accomplished using catastrophe theory.

(b) Dynamic inference from A_5 to A_3 and to $A_2(A_1)$.
In the following examples of archetypes taken from the A_k-series we can show that dynamic inference can be formalized by considering areas in a catastrophe diagram. If a larger area includes a smaller one, we can say that one of the corresponding semantic archetypes includes the other, i.e. the simpler archetype can be inferred dynamically. We arrive, thus, at a quasi-geometrical notion of inference.

In Fig. 6 we analyse subfields of the catastrophe diagram underlying the archetype of 'transfer'.

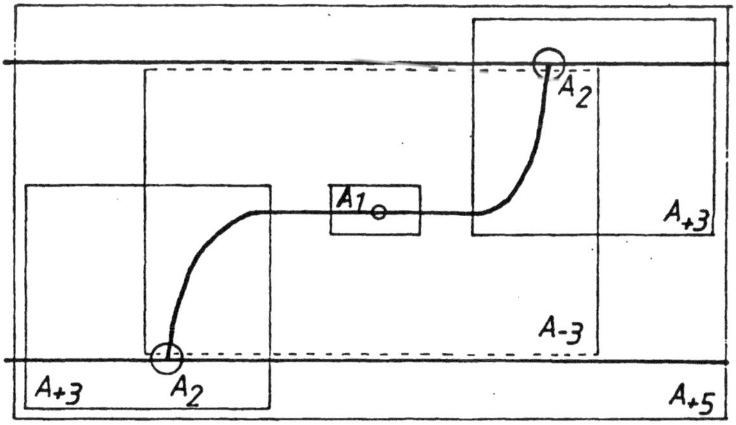

Fig. 6

We can distinguish four different types of fields:

(i) The whole field corresponds to a section in the butterfly (A_{+5}). The corresponding archetype was called the archetype of 'transfer'.

Example

Eve gives Adam an apple
M_1 M_3 M_2

(ii) In Fig. 6 we find two subfields of type A_3 (the cusp). They are the basis for two semantic archetypes: the archetype of emission and the archetype of capture

Examples

Eve gives an apple (emission)
M_1 M_2

Adam takes an apple (capture)
M_3 M_2

(iii) The attractor M_2 exists only between the two fold-points (A_2). Thus we have an archetype of delimited existence which was derived from the negative cusp (A_{-3}).

Example

The apple appears (in the scene) and disappears again.
M_2

(iv) The archetype of delimited existence was considered to be semi-elementary as it was a concatenation of two elementary archetypes derived from A_2 (the fold). In Fig. 6 we have two fold points - one where M_2 appears and one where M_2 disappears.

Examples

The apple appears (in the scene)
M_2

The apple disappears (in the scene)
M_2

(v) Finally all the points on the lines in diagram 5 which are not fold points are Morse, i.e. of type A_1 and structurally stable. For M_2 this stability is limited (cf. (iii)) by the fold points.

Examples

Adam exists
M_1

Eve exists
M_3

The apple exists (in the domain of M_2)
M_2

We assume that the relation of dynamic inference can be relevant in the semantics of verbs and in the description of tense, aspect and action modes. An elaboration of this application presupposes, however, a larger empirical and theoretical frame and cannot be given here.

The relevance of the type of semantics presented here to lexical semantics is immediate. The fact that in archetypal semantics we are seeking for primitive and irreducible mechanisms in the organisation of concept building and in the formation of semantic structure leads to direct applications in lexical semantics especially in the domain of fundamental relational and dynamic structures.

2. A Theory of lexical fields using catastrophe theory

In the past two rather similar but methologically contrary schemes have been used. The first which was motivated by the more holistic and dynamic methods in philology (and by the theory of fields in physics) is centered in the concept of the lexical field. The second method imitated procedures applied in phonetics and tried to decompose the words and morphemes into lexical components, features, scenes etc. Some authors tried to use this atomistic procedure to reconstruct the intuitive notion of a semantic field; compare Lutzeier, 1981, 1983). Compromises between a holistic procedure and componential analysis were sought by the theorists of semantic networks. Our aim is to renew the intuition behind field-theory and to propose adequate conceptual tools for it.

As Trier (1968:40), one of the fathers of the theory of lexical fields, already showed, we can take two different perspectives towards lexical fields.

(a) A lexical field is *disjunctive* (in Trier's terms), i.e. it is defined by its contour, its frontier, its membrane. This picture corresponds roughly to the idea of a mosaic put forward by Ipsen (1920).

(b) A lexical field is *injunctive*, i.e. it has a center and centripetal forces. Prototype semantics exploit this model of a field structure.
Fig. 7(a) and (b) give a picture of these notions.

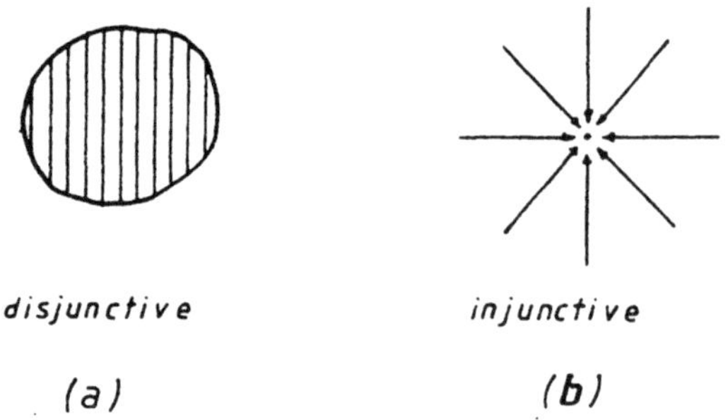

Fig. 7

Both pictures correspond to basic topological notions. In Fig. 7(a) we have the notion of a closed set, the 'membrane' is called the envelop of the set, in its environment we find points belonging to the set and not belonging to it. Fig. 7(b) gives a picture of a gradient-field with a punctual attractor and straight flows pointing to the center. As it stands the concept of disjunctive field is not provided with a dynamic. We can, however, define a similar notion in the domain of systems with Hopf-bifurcation and limit-cycles. In this case the contour is (ideally) a circle with attracting forces. Fig. 8 gives a picture of this situation. Similar dynamics play a role in thermodynamic systems.

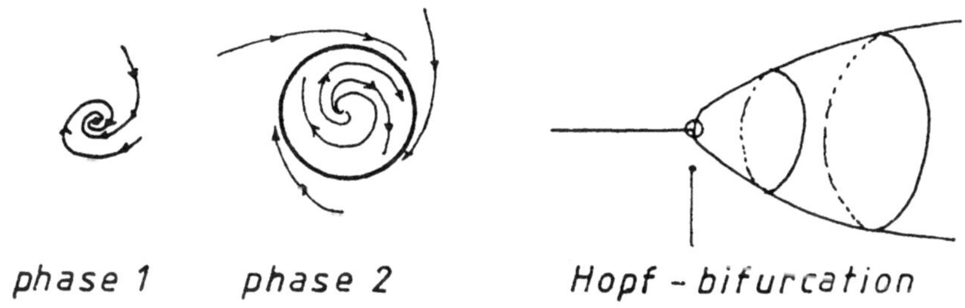

Fig. 8

Frontiers and membranes of the type shown in Fig. 7(a) can, however, appear also if more than one gradient-attractor exists in a field, i.e. if there exists a conflict of attractors. The conflict-lines or conflict-strata between different attractors define areas of attraction and instabilities, where the attraction changes. Catastrophe theory provides a classification for such basic instabilities and conflicts. It is in this line that our reconstruction of the theory of lexical fields will be pursued.

As a corollary we can say that in the domain of dynamic systems theory, other more sophisticated models are possible; C.T. models are the most fundamental and simplest possibility.

In Fig. 9 we show a rather complicated dynamic field taken from the center of the (compactified) elliptic umbilic. The picture shows a landscape; the altitudes are indicated by niveau-lines, in the same way as geographical maps represent a hilly district. We can see four basins B_1, B_2, B_3, B_4 and three hills H_1, H_2, H_3 (dotted lines). The basin B_4 is surrounded by three other basins and by three hills. The straight lines which intersect and form a triangle in the center, are saddles.

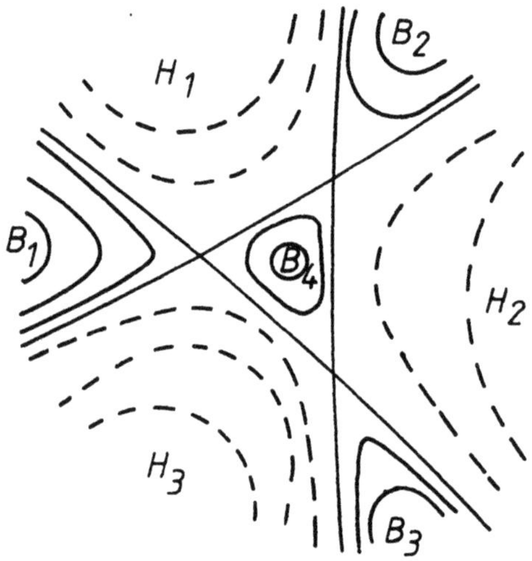

Fig. 9

In this case the saddles form a contour around B_4 and give frontiers to B_1, B_2, B_3. The vector field can be visualized if we reduce the number of inner variables to one, i.e. x. The corresponding situations are found in the cuspoid catastrophes. Thus for the cusp we find a conflict of two basins in the middle part of the unfolding, which in Fig. 10 is exemplified by a section though the bifurcation set (taking u as a negative constant).

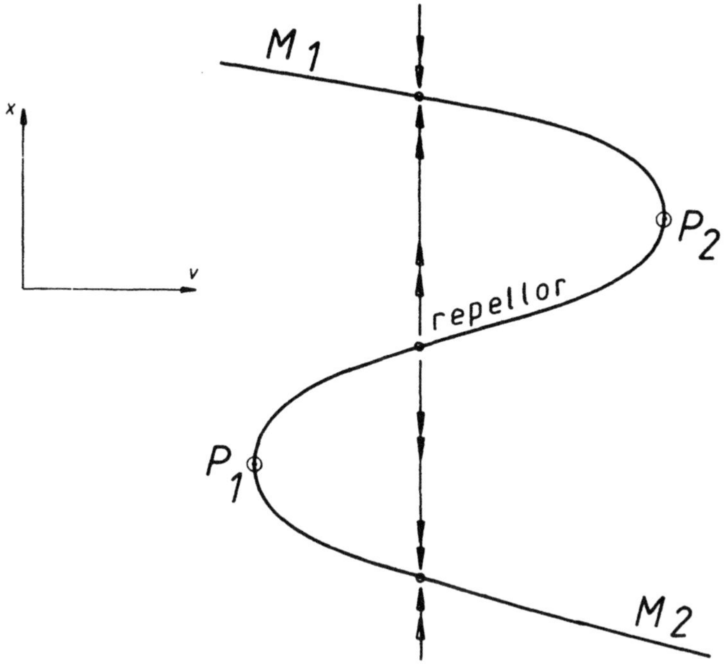

Fig. 10

Between the two saddles P_1 and P_2 we observe four vector flows, two attractors, M_1 and M_2, and one repellor, the line of maxima. If we take one constant value of v in the domain between P_1 and P_2, we can see the conflict of two basins of attraction.

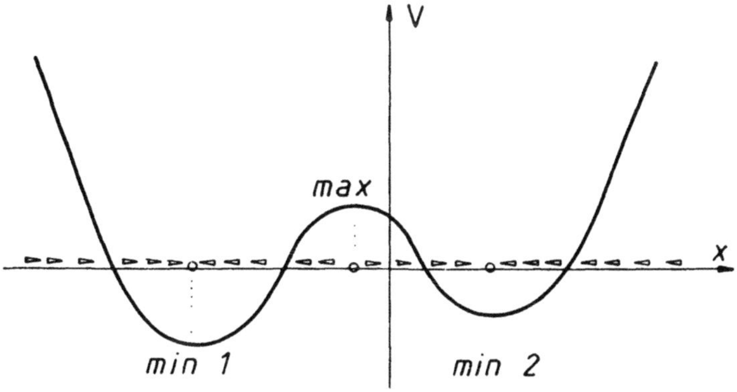

Fig. 11

If a ball is thrown into this field, we can imagine how it dances in the basin of M_1 or M_2 stabilizing finally in the attractor. If the movement is very strong the ball can, however, jump over the maximum and change its field of "residence".

In our semantic model the internal variable (x) is a semantic scale. The attractors corresponds to conceptual focuses,sinks of the field. With this stable center, the attractor, is associated a word. In the same way a field can subdivide a scale into two or three attracting domains, each domain can be subdivided into subdomains which correspond locally to the picture given in Fig.11. In elementary C.T. only these local phenomena are considered.

In a global model we can conceive other types of fields. We show some proposals made by Thom (1973a: 636f) and Thom (1980b: 26f). Fig. 12a shows the stable attractor type A_1; we can conceive an attractor "en falaise" with borderline as in Fig. 12b. The overhanging part of the basin leads to a frontier-catastrophe (with the possibility of hysteresis)

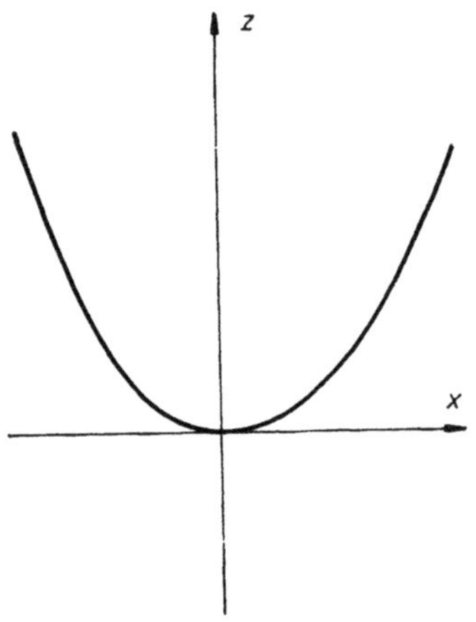

Fig. 12a

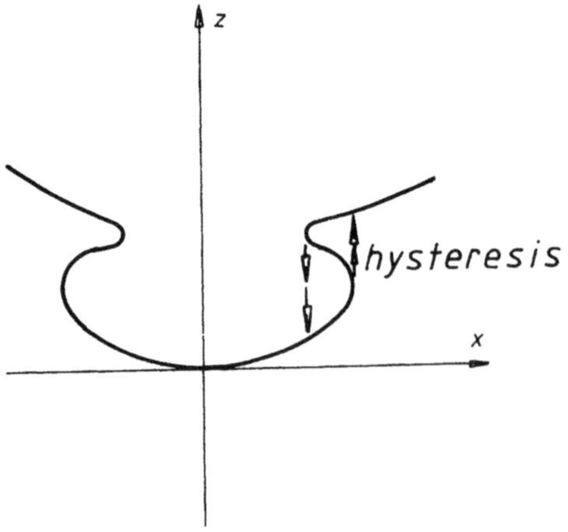

Fig. 12b

In a more general sense we can represent a field of concepts by a central basin, the prototype,and a surrounding field of lateral attractors. Fig. 13a gives a local picture, Fig. 13b looks from "above" on such a landscape.

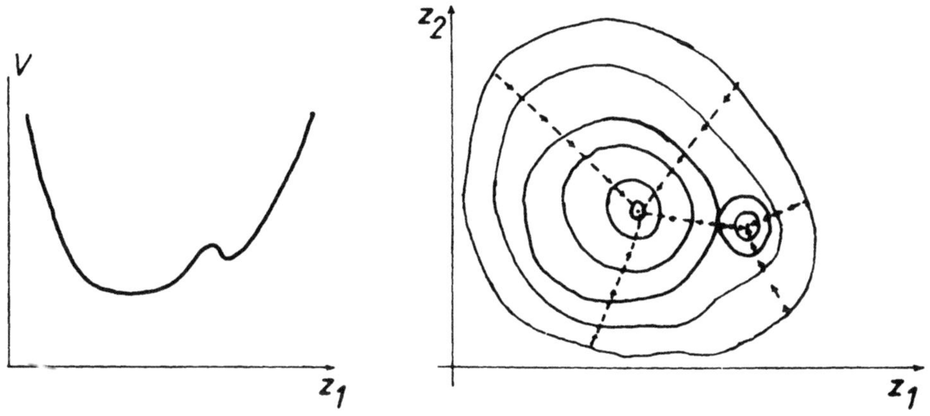

a b

Fig. 13

These morpho-dynamical landscapes can be very complicated globally (and are therefore far from the simplicity of C.T.-dynamics). It is only in their genesis that they are simple insofar as they are the result of many small processes which locally correspond to the archetypes described in C.T.-semantics. Fig. 14 gives a rough picture of a global field (we considered one axis only; in reality the number of dimensions can be high and complicated cross-dependencies can occur.

In the following sections we shall give an illustrative example for the application of notions introduced up to now. The main problem in the application of C.T.-semantics lies in a natural interpretation of the variables in the model or at least of the qualitative features of the model (frontiers, catastrophe jumps, polarities which appear and disappear etc.). If such an empirical

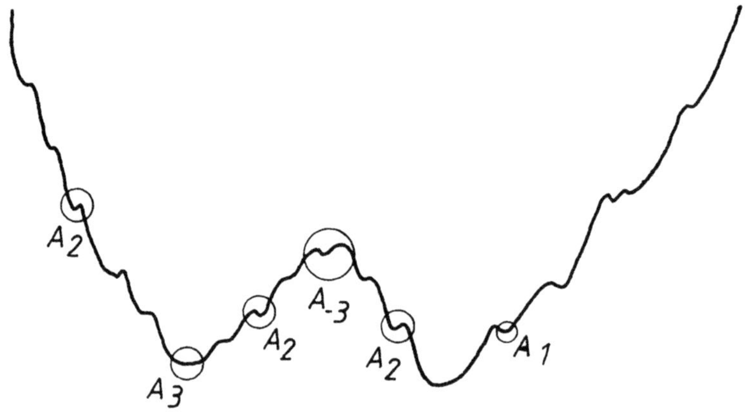

Fig. 14

interpretation is achieved it must be validated by the facts. As C.T. is a fundamental and local modelling device we must find fundamental domains of application and validation. Three areas of lexical semantics seem to fulfill this condition:

(a) The lexicon of colour terms (colour perception as background),

(b) The lexicon of evaluative adjectives (emotional dynamics as background),

(c) The lexicon of action-schemes.

2.1 The dynamics of colour perception and the lexicon of colours

We have good physiological models, good anthropological research done in many societies and a highly standardized research on the colour-lexicon of about 450 languages. In Wildgen (1981) we have described this field and the application of catastrophe theory to it in greater detail. We can summarize our results here (cf. Wildgen 1981 : 239-242, 285-288).

If we start from the opponent-cell theory of colour perception, we obtain two basic dynamic phenomena:

(a) The differentiation between excitation (presence of luminosity) and non-excitation. Dynamically we have an asymmetrical field with only one attractor (luminosity), which can be modelled qualitatively by the fold (A_2), cf. Fig. 15 .

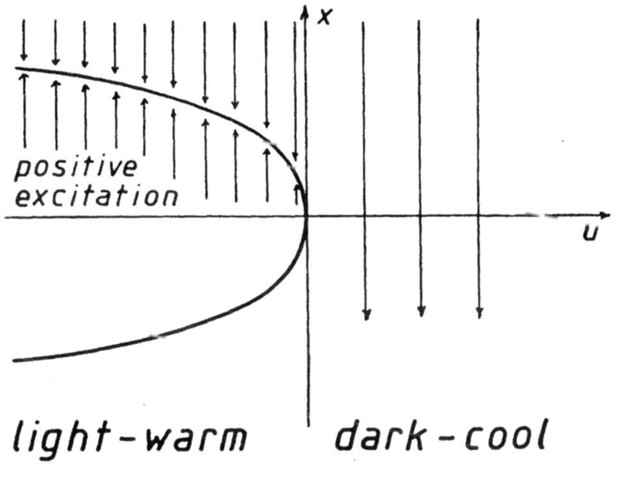

Fig. 15

(b) The further differentiation into the set of primary colours follows the principle of bipolar differentiation, which can be deduced from the cusp. First the domain specified positively in (a) is further subdivided. After the 'splitting' of light-warm into light (white) and warm (red-yellow), further cuspoid polarizations occur. We can thus reconstruct dynamically the results obtained by Kay and McDaniel (1978). Fig. 16 exhibits the sequence of cuspoid differentiation.

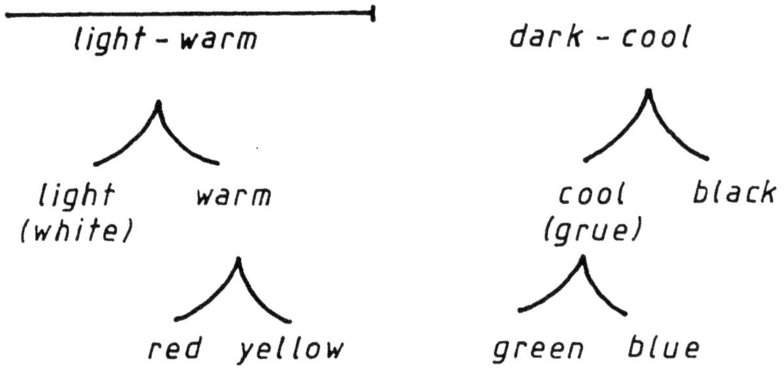

Fig. 16

Generalizing this fit between our deductive scheme and empirically found patterns in the differentiation of the lexicon, we can presume that there exists a perceptual basis for lexical semantics, which mirrors the dynamical organization of perception and which constitutes a repertoire of possibilities unfolded in the social history of a linguistic community.

2.2 The drive system and evaluative adjectives

Whereas in the case of colour perception the opponent-scheme is situated in cells on the retina (in the second layer after the cones, which react to three types of colours), the bipolar scheme for evaluative adjectives is deeper and older. It refers to our drive system and can be localized roughly in the brainstem (hypothalamus, subcortical kemels, parts of the reticular formation, cf. Bruter, 1976: 127-132). A behavioural correlate of this deep system is the bipolarity of attack-retreat, exemplified in the dog of Lorenz-Zeeman. Fig. 16 gives a picture of this situation showing the facial correlates of the emotional states 'fear' and 'rage'. The behavioural consequences are described by a cusp in Fig. 17.

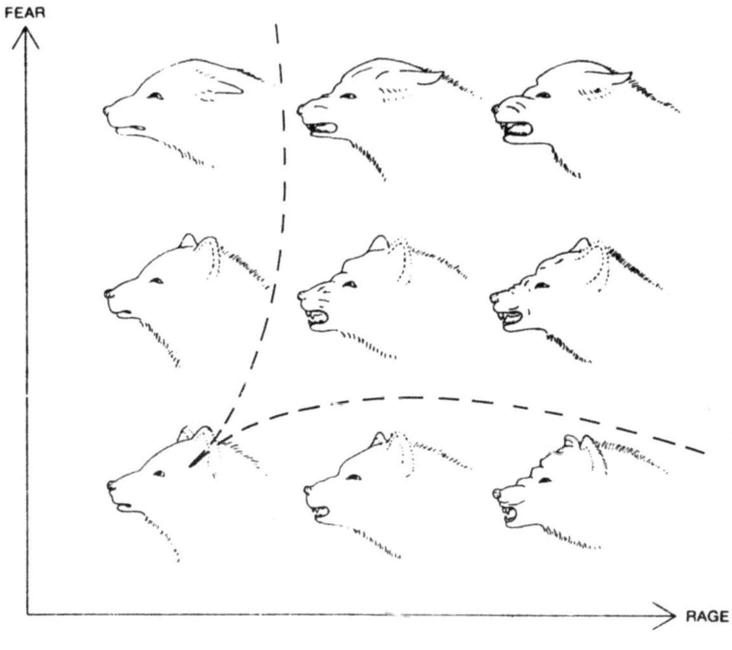

Fig. 17

The emotional split in a continuum of controlling parameters (fear, rage) is typical for processes in this domain and can be found also in the lexicon referring to emotional domains. The polarity principle found by Osgood and others is a consequence of this dynamic structure.

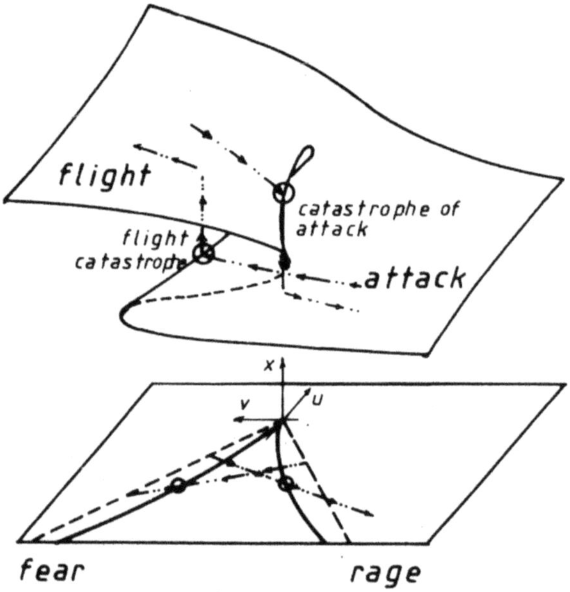

Fig. 18

Examples:

good	-	bad
beautiful	-	ugly
strong	-	weak

etc.

As a fundamental principle it is expanded to other domains, cf. big - small, long - short. Thus even scales which allow quantification are linguistically organized as qualitative fields (for further details cf. Wildgen, 1981). A generalization of our exemplary application could lead to a new formulation of differentiation theory in the glossematic tradition (cf. Lekomcev, 1974).

2.3 The dynamic classification of basic action schemes

In chapter 1.2 we characterized a set of semantic archetypes in which basic dynamic schemes found in catastrophe theory are interpreted as event and process schemes underlying natural language. This parallelism of abstract dynamic schemes and linguistic deep structures could be furthered by perceptual processes as Zeeman (1975) conjectured. Another hypothesis referring to psychoanalytically analysed processes of basic interactions between care giver and child has been elaborated by Mottron (1983). The hypothesis of Zeeman will be described here in order to complete the picture of possible interactions between perception and the structure of the lexicon (for Mottron's hypothesis cf. Wildgen and Mottron, forthcoming 1987: chapters 4 and 5).

A child may observe differences in the density of substance, i.e. in the distribution of relevant things in his/her environment. Centres of density (in terms of psychic relevance) are the own body, the body of the care giver and objects which can be transferred from one to another. In Fig. 19 six stages of such a transition T_0 to T_5 are presented schematically. A, B and C are centres of density (relevance). Fig. 19 shows that C is different from A and B as it is transitional, i.e. it is first held by A, appears in the scene, disappears in the scene, and is finally taken by B. In a concrete scene A could represent the body of the child, B the body of the care giver, C can be either food (from B to A) or a secretion of the child removed by the care giver (from A to B); in an later stage C can be a pet which partially replaces the care giver in his absence or it can stand for objects which are exchanged or pointed at and for other persons which are introduced to the child by the care giver (cf. Mottron's theory).

The archetype of transfer (cf. Fig. 5 and 6) could thus be discovered or developed by the child in the course of basic interactions and of subsequent perception of repeated interactions and of their storage in the memory.

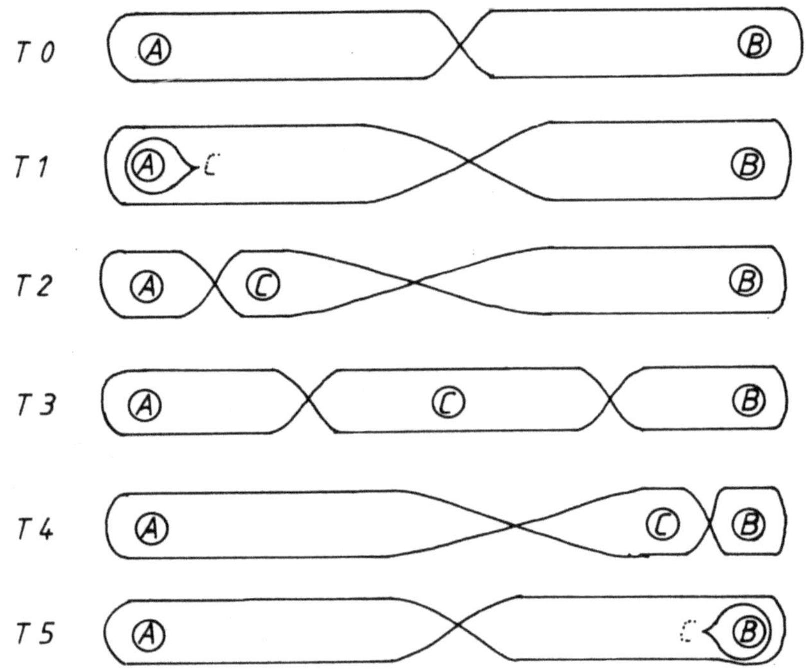

Fig. 19

If we consider more complex domains of the lexicon qualitative dynamics become insufficient as a modelling device. We reach a domain explored by Ch. Zeeman (1982) and which is constituted by three main components

(a) dynamic structure
(b) measure structure
(c) probability structure

(b) is in a way a connection between (a) and (c). In such a model the dynamic structure is considered as a preattentive mechanism in perception, the measure structure focusses attention and enables the observer/speaker to weight different features and to compute/ sum up the different aspects to a new holistic judgement. The probabilistic structure accounts for individual and social variations contained in all manifestations of the system. Thus/the model must aquire probability and measure structures to account for larger parts of the lexicon.

2.4 Processes of self-organization in the lexicon

The further development of the lexicon beyond the labelling of basic categories and scales can be considered as a self-organizing process which is determined by two main factors:

(a) the phonemic differentiation, which constitutes the material basis for morphemes and words,

(b) the semantic and contextual differentiation.

In Lindblom, MacNeilage and Studdert-Kennedy (1984) the basic mechanism of phonological differentiation is described as a process seeking an equilibrium between producer-based conditions (sensory discriminability, preference for 'less extreme' articulation) and listerner-based conditions (perceptual distance, perceptual salience) (cf. ibidem: 191). They show experimentally that a self-organizing process which obeys these conditions produces a phonological system and that we need no mental 'blue-prints' such as features, phonemes or the matrix of phonological oppositions. These are rather by-products of the self-organizing process.

A given phonological system allows the construction of a structured set of syllables or simple morphemes. We can distinguish two polar types of systems:

- a gestalt-like system in which the single morphemes cumulate several differences into morphemic "gestalts",

- an optimal phonological system, in which all oppositions are used such that no gaps are left.

Natural phonological systems lie between these extremes. Depending on a specific system of phonological differentiations we arrive at a basic set of syllables and simple morphemes.

The general structure of a lexicon is globally governed by quantitative laws which relate the length of words, their frequency (cf. Zipf's law), the measures 'polylexy', the number of readings per word) and 'polytexy' (the number of contextual fits per word) to the quantitative volume of the lexicon (cf. Köhler, 1985 and Altmann and Köhler, 1983). A currently directed research project by Gabriel Altmann: "Linguistic Synergetics" has as its objective to further develop this field of research. The mathematics used in Synergetics combine the qualitative theory of (nonlinear) dynamic systems and statistical dynamics (cf. Wildgen, 1986 for a first application to linguistics).

3. The construction of a semantic space as a theoretical framework of lexicology

3.1 The two dimensions of semantic space proposed by René Thom

The proposals made in the last sections aim at a rather narrow cooperation between physiological mechanisms and lexical structure. In this section I shall sketch some very global and rather speculative ideas, contained in recent papers by René Thom (Thom, 1978, 1980a, 1980b). Thom tries to give a coherent picture of the lexicon of human languages and of predication as a fundamental operation on the lexicon (cf. Thom, 1980b).

The fundamental structures of the lexicon can be projected on two main dimensions x and y. We shall give a short description of these dimensions.

(a) The fundamental dimension x has to do with the polarity: Speaker/hearer - world (as perceived, controlled, domain) or more general: ego - other, where 'other' is a cover term for other people and the world around ego. The extremum x = 0 is constituted by a kind of universal intersubjective space-time ($\mathbb{R}^4$).

In its neighbourhood, say at x = 1, we find subjective space-time, as it is represented in the mind of the speaker/hearer. As the fundamental conflict relating to this dimension is the one between subject (for itself) and object (in itself) we have two attractors separated by a threshold. This threshold is the domain of instabilities which lie at the heart of predication. The centres of simple autonomous sentences, mainly verbs and predicative adjectives, realize this intermediate domain and at the same time regulate the production of utterances.

The dimension x, called dimension: *object-subject*, allows for a rough classification of lexical domains. Figure 20 gives a picture of this classification.

The kind of lexical classes we find in many languages such as proper names, nouns, verbs, adjectivs, pronouns, conjunctions can be projected on the dimension x. The conjunctions, especially logical conjunctions seem to be almost free of "empirical", "descriptive" content; other particles like deictics take their content from the situation of use and from nonverbal actions co-present with them (the pointing of the finger etc.). When the child first makes two-word utterances, we encounter sentences consisting of nominal (substance) words and deictics as in *papa here*. This fundamental bipolarity is overcome by the development of relational and dynamic concepts: verbs and predicative adjectives. The forming of a sentence can be understood as a dynamic moving on the scale x or resolving conflicts on this scale (at least in the initial phase of sentence building by the learner,

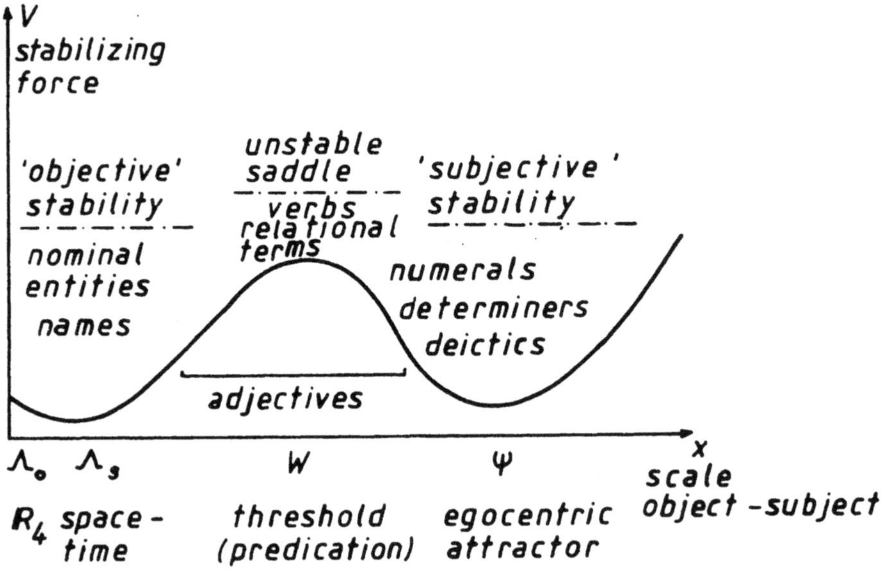

Fig. 20

later these dynamics get ritualized). The term 'concept' can in a strict sense only be applied to nominal entities (in the objective attractor). The other entities need for their stabilization the activity of the speaker/hearer which is immersed in space and time and thus changes rapidly with situations. This is especially true for deictics, but it extends to verbs and adjectives (some of the adjectives analyzed in the last section especially adjectives of colour and shape are on the slope towards the objective attractor and therefore in the neighbourhood of nominal entities)*.

(b) The second dimension, y, is called *semantic complexity*. A concept A is more complex than a concept B if its geometrical domain is more narrow, if the associated neuro-physiological processes are more specific or if the domain of a quality space is more restricted (cf. Thom, 1980b: 30 for a definition). This dimension can be compared to Carnap's notion of "semantic information" (cf. Wildgen, 1977, chapter 3.5.).

As the conceptual structure and its stability depend on the axis object - subject, it follows that the values of semantic complexity decrease with growing values on the x-axis. Full semantic variability (range of values on the y-axis) can be found in the lexicon of nominal entities. The following list gives a first view at the sortal richness on the y-scale (from maximal values to minimal ones).

A. *Human beings*

Concepts relating to human beings contain several layers, to which we have objectively or intuitively access: their social behavior, their movements, their emotional states and attitudes, their belief, their knowledge etc.

---

* Thom says (in Thom, 1980b: 27f) (translation by the author): "A mayor difficulty of every linguistic theory lies in the fact that the notion of concept is not sufficient to take account of the elementary units of linguistic activity; the concept is in fact only suited to describe these units corresponding to one grammatical category, the noun. Most of the other categories call for another description; their semantic stability makes necessary an activity (muscular or mental) of the speaker and this activity is immersed in variable ways into space and time. For example, the meaning of a deictic, i.e. its spatial reference, depends in essential regards from the linear prolongation of the demonstrative gesture. But even the concept has a stable meaning only in a very first approximation."

B. Animate beings interacting with human beings: We have a good idea of their form in space (and time), if the animal belongs to our environment; we have some access to its behavior, but its inner activities are more or less obscure to us (we can expand our introspection to gain hypotheses on the inner processes of animals in the same way such expansions are used to understand other persons; the degree of reciprocal control on these expansions is, however, very low).

C. Living beings which do not interact with humans. We only have access to their form in space and in time (their 'life-span').

D. Artefacts. The objects produced by men are in a certain sense emanations of human beings. They are created and cast away/lost (i.e. they have an artificial 'life-span'), they can do specific things and they can even replace animals and human beings in certain functions. (Since the eighteenth century the vision of quasi-human robots has stimulated human imagination.)

E. Natural objects. The normal perspective focusses on their form in space and on their use in human contexts.

F. Primary matter and primary qualities. They have no form in space and time; objects are made out of primary matter (in a human perspective) and subjective impressions are caused by qualities.

Two further dimensions are hidden, behind the dimension 'semantic complexity'. These are 'semantic contiguity' and 'metaphor' (cf. Thom, 1980a:201, where these notions are taken from Jakobson and related to the diffusion of meaning in the lexicon). These dimensions are rather language - and user-specific and establish a gradient called 'gradient of necessity' (cf. Mottron 1983:18-55). Roughly the semantic content, in Thom's terms the 'prégnance', diminishes along the path of metonymical and metaphorical expansion.

In Fig. 21 the correlation of the two basic dimensions is shown (cf. Thom, 1980b:30)

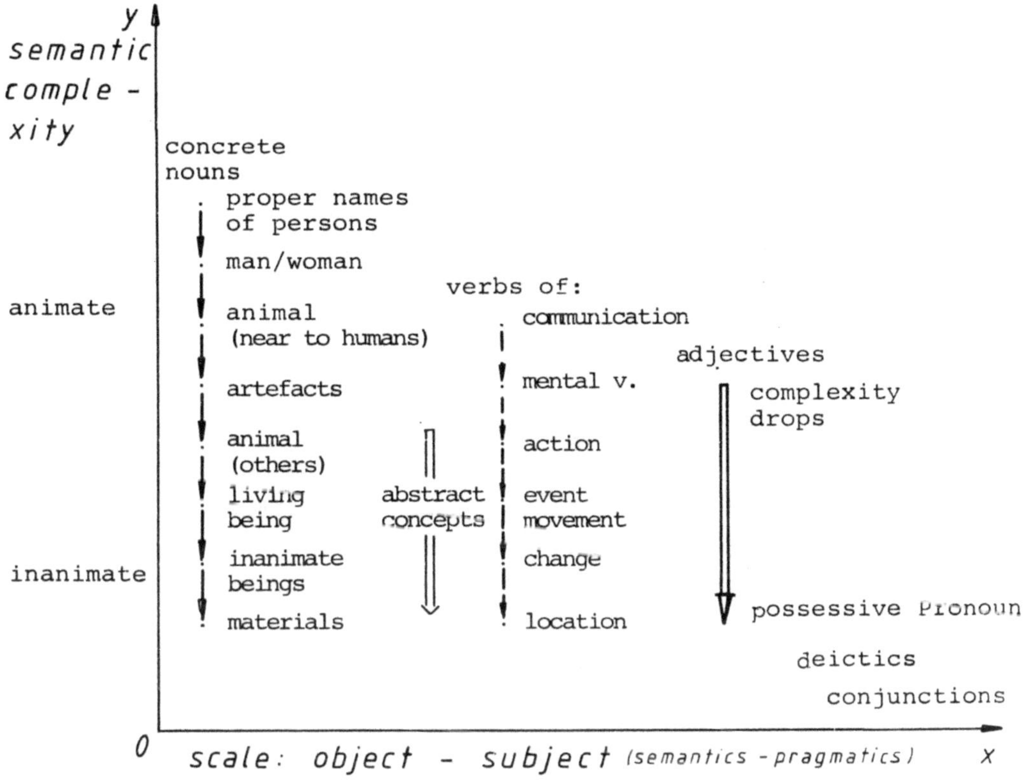

Fig. 21

In a sense the scale 'object - subject' leads secundarily to a methodological scale: semantic-pragmatic. Nouns are nearer to concepts (cf. above) and easier treatable with the traditional methods of semantics (i.e. componential analysis, lexical field analysis). Towards the end of the scale only methods using contextual, userspecific informations can describe the 'content' of lexical entities. A special kind of particles has only organizational functions, they are 'syntactic' in a wider sense (conjunctions, markers of quantification and scope, anaphora etc.). They are better treated in text-semantics and text-pragmatics and are outside the scheme in Fig. 21 which takes account only of entities and processes in the clause (the semantically autonomous and minimal entities, cf. Thom, 1980b:27).

The lexicon is part of the longterm consequences of the unfolding of our cognitive universe insofar as it is organized / represented by linguistic structures. Syntax can be conceived of as the outcome of fundamental regulations of the lexical unfolding and as a further application of the basic principles which govern this morphogenesis, however, there is more to it than that because the development of syntax chooses between the many possible products of the morphogenetic process which is only well determined in its first stages. A further restriction concerns the socio-historical dynamics which form our language. The child is confronted with this historical process as soon as it "learns" the language of adults. A theory of language and grammar must take into consideration the historical, social and spatial development of linguistic communities, if the theory is to be fully explanatory. Some aspects of the historical dynamics of the lexicon will be touched in the following section.

3.2 A catastrophe theoretic analogue of the concept of a three dimensional semantic space proposed by Ballmer and Brennenstuhl

Since the mid-seventies Thomas Ballmer and Waltraud Brennenstuhl analysed the lexicon of German verbs, adverbs and nouns. Applying very general classificatory principles (presupposition and similarity) they arrived at a fundamental structure which organizes the lexicon of verbs: the three dimensional semantic space (cf. Ballmer, this volume). The three dimensions are the following:

(a) The type of process ('aktionsart'); its beginning, its central phase, and its ending.

(b) The intensity of the process (highest in the central phase).
(c) The (evolutionary) level of the process, its complexity.

Although some analogies to Thom's dimensions are evident, the framework is different and we must go beyond catastrophe theoretic semantics to reconstruct the semantic space of Ballmer and Brennenstuhl within our framework. This undertaking elaborates our model and provides a mathematical foundation to the empirical classification proposed by Ballmer and Brennenstuhl.

ad a: The dimension, type of process, denoted by the verb: "aktionsart"

Thomas Ballmer (1982a:513) characterizes this dimension as follows:
"the *aktionsart* is a measure of the development of a process (beginning, ongoing, ending; or in linguistic terms: inchoative, durative, terminative). The aktionsart correlates with the (onto) genesis of objects, organisms, actions, etc. (...) Along the aktionsart are situated the closed processes with a natural beginning, end, ..."

Within the framework of our model there are the following three phases which correspond to the 'aktionsart':

- the archetype of birth/beginning,
- the archetype of death/ending,
- the archetype of steady process without catastrophe (the 0-unfolding).

The typical processes underlying Ballmer's "hat structure" is found in the negative cusp (A_{-3}). Fig. 22 shows the bifurcation lines of the negative cusp. The minimum appears in the inner zone, beneath the singular point A_{-3}.

In the light of archetypal semantics the two bifurcation catastrophes death/birth are basic, but the specific combination used

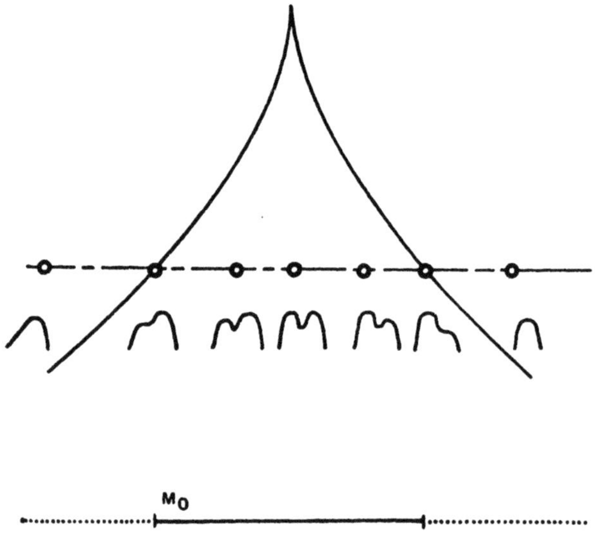

Fig. 22

in Ballmer's model is only one element in a larger list of process types. Furthermore semantic archetypes can show interactions of maximally four nominal rôles (five if the double cusp is considered).

If the singular events between the agents are linearily ordered, we can represent them in one dimension (losing the space character of the interaction). We shall introduce a notation for archetypal events which preserves the linear representation proposed by Ballmer and Brennenstuhl. We distinguish three types of basic phenomena:

a) *Bifurcation events:* birth/beginning and death/ending. They are noted as:

   b ├──────── birth; ────────┤ b death

b) *Conflict catastrophes*. They are noted as: c
In these points dominance changes from one regime (attractor) to another. Both regimes coexist in the neighbourhood of these points.

c) The domains of stability of a regime R_i are noted above the line:
example: R_i

With these conventions we can represent the principal process types found in the A-series of catastrophes (the cuspoids):

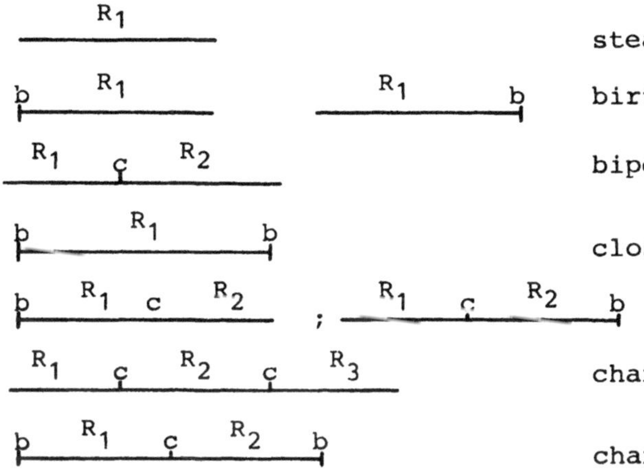

A_1	steady processes;
A_2	birth/death processes
A_3	bipolar change
A_{-3}	closed life span/existence
A_4	change and death/birth
A_5	changes in a three modal field
A_{-5}	change in a closed domain

A_6, A_7, A_{-7} ... continue this typology in a linear way.

Table 2

The characteristic "hat structures" proposed by Thomas Ballmer can be elaborated on the basis of the richer typology of processes shown in Table 2. We give two examples related to the positive (A_5) and the negative butterfly (A_{-5}) in Fig. 23.

This shows schematically the enrichment of Ballmer's "hat structures" in our analogue. Compared to catastrophe theoretic semantics we have lost some information, i.e. the spatial configuration

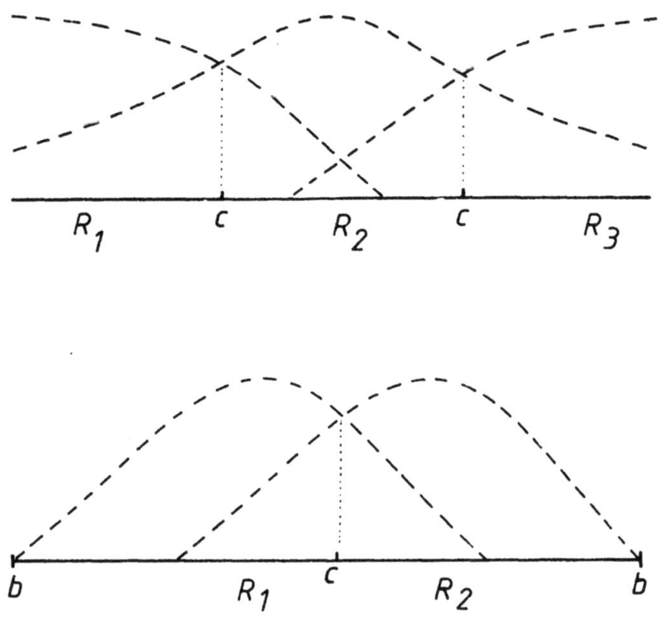

Fig. 23

of the agents in a scene (qualitatively) and the quick dynamics of the stabilizing or destabilizing vector flows.

ad b: The dimension: activity intensity

Ballmer's definition of this dimension is as follows:
"the activity intensity is a measure of the actual activity of the subject in the process labeled by the verb). At the beginning of the process the activity is low, in the middle phase there is a maximum of intensity, towards the end it falls again." (ibidem: 513). This dimension at the same time correlates to the number of verbs included in classes of verbs denoting similar processes. Ballmer explains this correlation by the principle of relevance which says that relevant features are more prominent in linguistic expression. He states:
"Because there are antropologically more relevant meanings than others it does not come as a surprise that the differences in lexi-

calization, length, formal and phonetico-phonological simplicity, frequency of occurrence in a text, etc. are related to the intrinsic shape of the semantic space. The more relevant a meaning, the more succinct the expression (this is a brief version of what we call the *principle of relevance*)" (Ballmer, 1982b: 491).

The three features:

(a) the intensity of the process or rather of its single phases,

(b) the frequency of lexical items standing for these phases of the process, and

(c) the linguistic/cognitive/social relevance of the phases denoted,

map onto one dimension.

We shall try to explicate the correlation between the different features contributing to the measure called 'activity intensity' and to integrate it into our framework. The following assumptions could be an appropriate starting-point:

(a) Singular points, i.e. bifurcation points (b) (beginning, ending of a regime) are dynamically very intensive. Thus we can presume that the number of *different* labels is minimal and that the type/token ratio is low for the lexical realisation of these process phases.

(b) Regular points and their environment (the stable phases of the process) can be maximally differentiated using secondary non-processual features. As a consequence the type/token ratio is higher.

If these assumptions are correct, we get two different measures which stand in a complementary relation to each other: a function for the intensity of the different phases of a process, and a function for the amount of lexical differentiation. The curve of lexical differentiation would correspond to the "hat structure" advocated by Thomas Ballmer. In our model it shows a Gauss-distribution of lexical labels for different phases of a process (type A_{-3} in catastrophe theory).

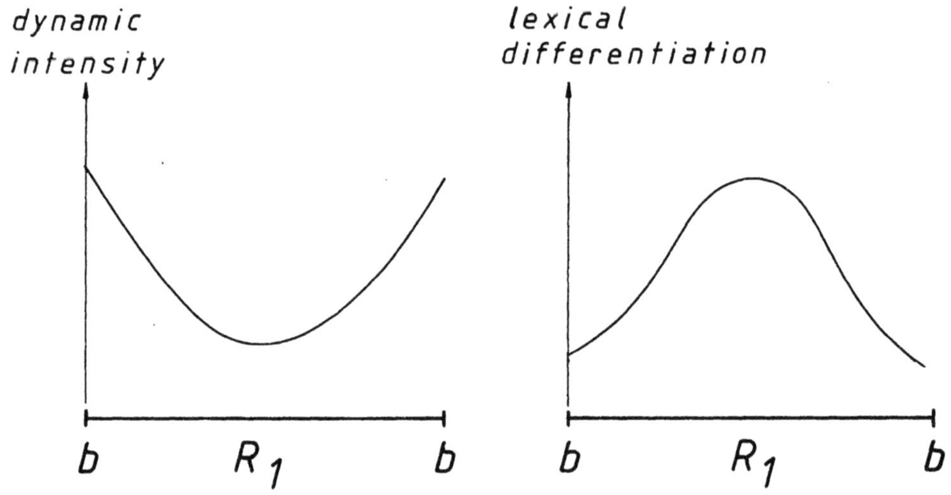

Fig. 24

In our elaboration of the dimension called "aktionsart" by Thomas Ballmer we introduced a new type of dynamic structure not contained in the model proposed by Ballmer and Brennenstuhl: conflict catastrophes. In our context (cf. the assumptions above) it is plausible that the relevance of conflict phases is very high and that if lexical differentiation is a function of the internal process structure, it grows in the environment of conflict catastrophes. In the simplest case we would arrive at an additive function of the two regimes which are in conflict. Fig. 25 shows the graph of the function "lexical differentiation" on this assumption.

The measure for lexical differentiation increases above the level of the stable phases M_1 and reaches a maximum in the environment of the conflict point c.

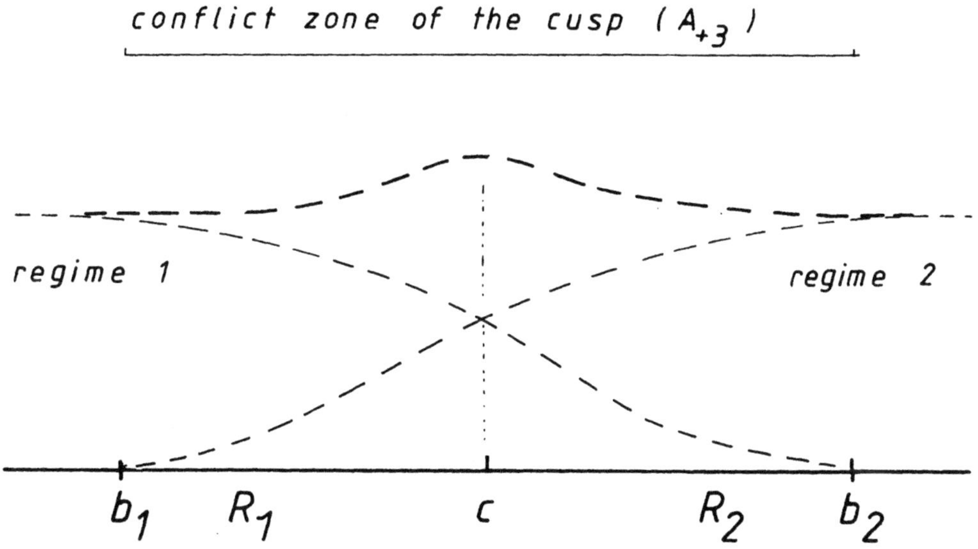

Fig. 25

The above mentioned assumptions are probably only rough approximations, but they correspond to the observations made by Ballmer and Brennenstuhl in the context of their analysis of German verbs (cf. also Ballmer and Brennenstuhl, 1985). A further elaboration of the model is likely to lead to lexicostatistic predictions which can be tested empirically.

ad c: The dimension called "eingriffsgrad", i.e. degree of impact

Ballmer (1982a: 513) defines the dimension: "eingriffsgrad" as the: "measure of fugitive or resistant survival potential against attacks upon the object/organism in question. The eingriffsgrad correlates with the process of bio-physical evolution (cf. Ballmer, 1981b) and neuro-physiological morphology".

If we take the maxima on the scale called "activity intensity", i.e. lexical differentiation for groups of verbs, as the vertical axis and the (evolutionary) "eingriffsgrad" as the horizontal axis, we obtain "shells" which are internally subdivided by "models" (smaller groups of verbs). Fig. 26 shows the series of "shells" a, b, c on the axis "eingriffsgrad".

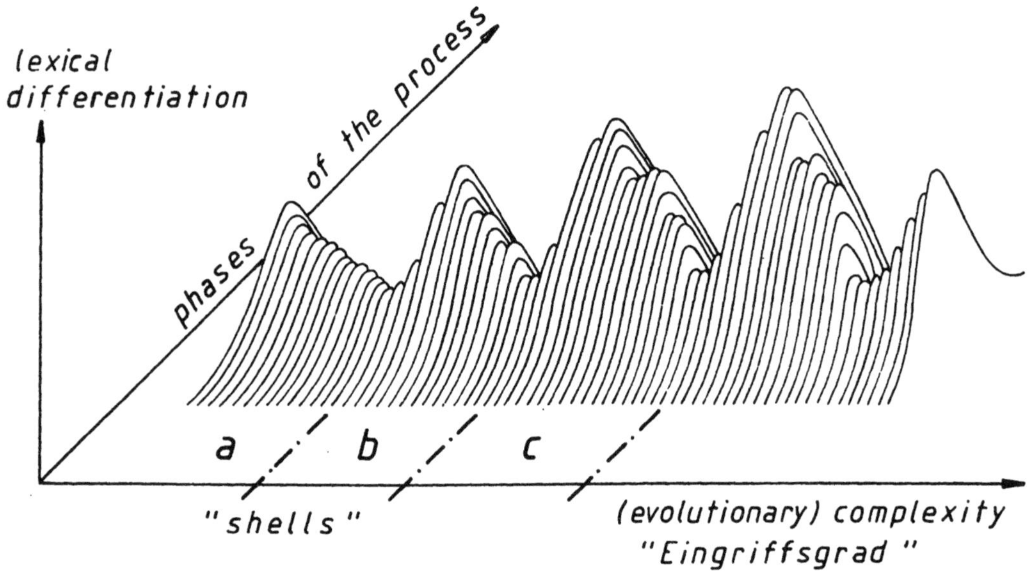

Fig. 26

The main groups of shells and subgroups found by Ballmer and Brennenstuhl in the lexicon of German verbs are correlated with basic types of sentences; they are shown in Table 3 (cf. Ballmer, 1982a: 499f).

The discontinuities in the semantic space (cf. Fig. 26) mirror typical discontinuities produced by evolutionary processes. These ontological discontinuities partially correspond to discontinuities

shells / subgroups			sentence types
1	*being*	(nonactive)	intransitive
2.	*influence*	(active)	
2.1.	action		
2.2.	grasp		transitive
2.3.	modification		
3.	*communication*		hypertransitive

Table 3

in our cognitive system and are secondarily inherited by the linguistic system. In this sense the lexicon can be seen as an image of the world and of the human mind.

This deep hypothesis can be traced back to Goethe's thinking on morphology or even to Aristoteles. It comes very near to Thom's philosophy of nature in its application to the mind and to language. In a certain sense it is opposed to the hypothesis of linguistic relativity which originated probably in sophistic philosophy (homa mensura) and was revived by Humboldt and later by Sapir and Whorf (cf. Wildgen, 1985b).

4. Conclusions

A theory of the lexicon which is explanatory in the sense that it uses very fundamental laws of nature, of our biological heritage and our mind can be developed in the framework of the dynamic paradigm in linguistics. It is connected to recent advances in mathematics, in theoretical biology, psychology, phonetics and lexico-statistics. It leads to interesting hypotheses which can be tested empirically. Although the formal nature of the theory should be further clarified, it is of eminent importance to begin new empirical research on the basis of the hypotheses put forward within the dynamic paradigm.

References

Altmann, G. and R. Köhler, 1983 "Systemtheorie und Semiotik". In: *Zeitschrift für Semiotik*, 5, pp. 424-431.

Arnol'd, V.I., 1972 "Normal Forms for Functions near Degenerate Critical Points, the Weyl Groups of A_k, D_k and E_k and Lagrangian Singularities. In: *Functional Analysis and its Applications*, 6, pp. 254-272.

Ballmer, Th., 1982a, "The Roots of ἐπιστήμη: Archetypes, Symbols, Metaphors, Models, Theories". In: *Poetics* 12, pp. 493-539.

Ballmer, Th., 1982b, "The Interaction between Ontogeny and Phylogeny: A Theoretical Reconstruction of the Evolution of Mind and Language." In: Koch, W.A. (ed.) *Semiogenesis*. Essays on the Analysis of the Genesis of Language, Art and Literature. Lang, Frankfurt/Bern, pp. 481-545.

Ballmer, Th., 1987, "Case, Aktionsart and Ergativity. A Semantical Base for Typological Issues." This volume.

Ballmer, Th. and Waltraud Brennenstuhl, 1985, *Deutsche Verben*. Narr, Tübingen.

Bruter, C.P., 1976, Topologie et signification, vol. 2: Aspects neurophysiologiques. Mailoine, Paris.

Eikmeyer, H.-J. and H. Rieser (eds.), 1981. *Words, Worlds and Contexts*. New Approaches to Word Semantics. De Gruyter, Berlin.

Harvey, I.H. et alii (Eds.), 1971, *New Directions in Attribution Research*. Erlbaum, New Yersey.

Gilmore, R., 1980, *Catastrophe Theory for Scientists and Engineers*. Wiley, New York.

Heider, F., 1958, *The Psychology of Interpersonal Relations*. Wiley, New York.

Ipsen, G., 1924, "Der alte Orient und die Indogermanen". In: Streitberg, W. (ed.), *Stand und Aufgaben der Sprachwissenschaft*. Heidelberg, pp. 200-237.

Jones, E.E. et alii, 1971, *Attribution*: Perceiving the Causes of Behavior. General Learning Press, Morristown.

Kay, P. and C.K. Mc Daniel, 1978, "The Linguistic Significance of the Meanings of Basic Color Terms". In: *Language*, 54, pp. 610-646.

Köhler, R., 1985. *Die Selbstregulation der Lexik,* Dissertation, Bochum.

Lekomcev, Ju.K., 1974, "The Glossematic Theory of Linguistic Oppositions and Differentiation Theory in Semantics and Descriptive Semiotics". In: *Linguistics,* pp. 37-72.

Lutzeier, P.R., 1981, *Wort und Feld.* Wortsemantische Fragestellungen mit besonderer Berücksichtigung des Wortfeldbegriffes. Niemeyer, Tübingen.

Lutzeier, P.R., 1983. "Wortfelder als Maßstab für Interpretationen am Beispiel des Feldes der Stimmungen im Deutschen." In: *Zeitschrift für Sprachwissenschaft,* 2, 45-71.

Mottron, L., 1983, *Contraintes communes à l'acquisition, la théorisation et la pathologie de la deixis,* Thèse d'Etat. Sorbonne, Paris.

Poston, J. I.N.Stewart, 1978, *Catastrophe Theory and its Applications.* Pitman, London.

Rieser, H., 1985, "Prozedurale Grammatik und Wortsemantik." In: Rieger, B. (ed.) *Dynamik in der Bedeutungskonstitution.* Buske, Hamburg, pp. 19-30.

Thom, R., 1973, "De l'icone au symbole: esquisse d'une théorie du symbolisme". In: Cahiers Internat. du Symbolisme, 22/23, pp. 85-106.

Thom, R., 1977, *Stabilité structurelle et morphogénèse.* Essai d'une théorie générale des modèles. (Second, corrected and augmented edition; first edition 1972). Interéditions, Paris.

Thom, R., 1978, "La double dimension de la grammaire universelle." In: Seiler, H. (ed), *Language Universals.* Narr, Tübingen, pp. 79-87.

Thom, R., 1980a, "L'espace et les signes." In: *Semiotica,* 29, pp. 193-208.

Thom, R., 1980b, "Prédication et grammaire universelle." In: *Fundamenta Scientiae,* 1, pp. 23-24.

Trier, J., 1968, "Altes und Neues vom sprachlichen Feld." In: Schmidt, L. (ed), *Wortfeldtheorie.* Darmstadt, 1973, pp. 129-161.

Wildgen, W., 1977a, *Differentielle Linguistik.* Entwurf eines Modells zur Beschreibung und Messung semantischer und pragmatischer Variation. Niemeyer, Tübingen.

Wildgen, W., 1977b. *Kommunikativer Stil und Sozialisation.* Eine empirische Untersuchung. Niemeyer, Tübingen.

Wildgen, W., 1979. *Verständigungsdynamik*. Bausteine für ein dynamisches Sprachmodell, Habilitation, Regensburg, Ms. (partially published in Wildgen, 1985a).

Wildgen, W., 1981, "Archetypal Dynamics in Word Semantics: An Application of Catastrophe Theory." In: Eikmeyer, H.J. and H. Rieser (1985), pp. 234-296.

Wildgen, W., 1982, *Catastrophe Theoretic Semantics*. An Elaboration and Application of René Thom's Theory. Benjamins, Amsterdam.

Wildgen, W., 1983, "Goethe als Wegbereiter einer universalen Morphologie (unter besonderer Berücksichtigung der Sprachform). In: *Jahresbericht des Präsidenten der Universität Bayreuth*, 1983, pp. 237-277. Reprinted in: L.A.U.T. Preprints, Series A, No. 125, Trier.

Wildgen, W., 1983, "Modelling Vagueness in Catastrophe Theoretic Semantics." In: Ballmer T.T and M. Pinkal (eds.), *Approaching Vagueness*. North-Holland, Amsterdam, pp. 317-360.

Wildgen, W., 1985a, *Archetypensemantik*. Grundlagen für eine dynamische Semantik auf der Basis der Katastrophentheorie. Narr, Tübingen.

Wildgen, W., 1985b, *Dynamische Welt- und Sprachauffassungen* (in ihrer Entwicklung von der Antike bis zur Gegenwart). Zentrum Philosophische Grundlagen der Wissenschaften, Bremen.

Wildgen, W., 1986, "Synergetische Modelle in der Soziolinguistik. Zur Dynamik des Sprachwechsels Niederdeutsch - Hochdeutsch in Bremen um die Jahrhundertwende (1880-1920)." In: *Zeitschrift für Sprachwissenschaft*, 5, pp. 105-137.

Wildgen, W. and L. Mottron, forthcoming 1987, *Dynamische Sprachtheorie*. Sprache als selbstorganisierendes System / Morphogenese der Bedeutungsstruktur, Brockmeyer, Bochum.

Zeeman, Ch., 1975. "Catastrophe Theory. A Reply to Thom." Reprinted in: Zeeman, Ch., 1977. *Catastrophe Theory*. Selected Papers 1972-1977. Addison-Wesley, Reading (Mass.), pp. 622 -632.

Zeeman, Ch., 1982, *Multistable Perception*. Paper presented at the Conference on Logos and Catastrophes. Centre Culturel de Cerisy, September 1982 (the Proceedings are forthcoming).

DYNAMIC ASPECTS OF NOMINAL COMPOSITION

Wolfgang Wildgen
University of Bremen

Nominal compounds constitute a phenomenon which is important in several respects:

(a) it is an important means of enrichment for our lexicon;
(b) it can replace larger structures and thus serve as a strategy for language economy;
(c) the creation and interpretation of novel compounds makes use of contextual and encyclopedic knowledge, thus introducing pragmatic factors.

A fourth aspect, namely the typological prominence of composition in, for example, German and English and its absence in other languages will be mentioned in the last chapter.

In the description and modelling of nominal compounds, three perspectives can be distinguished:

(a) The local perspective. Here only the constituents of the compound, their syntactic and semantic properties and the constituent structure of the compound are taken into account. The local perspective is more or less a methodological fiction which helps to analyse basic processes. These are called *microprocesses*.
(b) The global perspective. The novel compound is embedded in a phrase, a sentence, a text. It is related to preceding and following structures in the text. If we examine those processes relating the compound to features of the text (or even the dialogue) we arrive at *macroprocesses* governing the creation and interpretation of compounds.
(c) The analogical perspective. A special context is consti-

tuted by the linguistic knowledge of the speaker and the hearer. A novel compound can be related to a known compound, to a class of such compounds or even to a general regularity by compounds with a specified micro-structure.

These perspectives are complementary rather than exclusive. The following chapters summarize results of empirical work done by the author within the framework of the Regensburg project "Aktualgenese nominaler Komposita". More detailled information can be found in Wildgen (1980, 1981a, 1981b, 1982b).

1. Local processes in nominal composition

In traditional treatments of nominal compounding, one point of departure has been the syntactic and morphological category of the constituents (cf. Fleischer, 1969). For the semantics of nominal compounds we must, however, choose more fundamental categories. We consider this to be achievable by separating nominal, static entities from relational (verbal), dynamic entities. This distinction has already been shown to occur in the phase of one-word-sentences during language-acquisition as a first functional differentiation (cf. Bloom and Lahey, 1978: 110-113, and McCune-Nicholich, 1980). We hypothesize that a nominal compound must normally contain at least one nominal, static constituent and one relational, dynamic constituent. In those compounds where this basic functional pattern is not realized, information may be extracted from one (or both) of the constituents. Thesis 1 expresses this hypothesis.

Thesis 1

The basis for the semantic interpretation of nominal compounds is a bipolar field constituted by a static, nominal entity and a dynamic, relational (verbal) entity. Thus nominal compounds tend to activate semantic structures which are early in an ontogenetic and phylogenetic sense. It follows that compounds are not derived in a strict sense from more elaborate phrasal or syntactical constructions.

Thesis 1 seems to contradict the transformational treatment of compounds, which has motivated much of the recent work in this field. This contradiction is, however, moderated by the fact that the pattern of nominal compounds is part of a *polycentristic field* (cf. Dressler, 1977), such that similar structures appear at the levels of word formation, syntactic phrases and sentences. The levels are not autonomous as postulated by structuralistic methods but make a selection from among recurrent patterns. Thus the basic intuitions of transformationalist research can be saved without adopting its rather restrictive framework.

Nominal compounds belong to the level of words, near the boundary between word structures and phrasal structures. In a dynamic perspektive, the boundaries of structural levels can be compared to semi-permeable membranes, which control structure maintenance and diffusion. Inside the "word-membrane" there exist certain criteria of completeness versus deficiency. In thesis 1 the basic regularity is indicated. If a noun contains two stems (if it is a nominal compound) it must have a relational kernel and a nominal filler. Contrary to the completness criteria at the syntactic level, the arguments of the relational predicate (= the attractors of the dynamic centre, cf. chapter 2) can be selected using a measure of prominence.

If this normal structure is not present in a nominal compound, we can predict a flow of structural information through the membrane. Thesis 2 formulates this consequence.

Thesis 2

If the compound is functionally deficient we can predict a flow of structure diffusion[1]:

(a) from the level of word semantics to the level of word

1 The term "diffusion" has gained a new theoretical dimension in mathematical biology and ecology; compare Okubo (1980), Rubinow (1980, lecture 9).

structure, i.e. the internal semantics of the constituents is exploited;

(b) from the level of context to the level of word structure, i.e. thematically similar structures in the context are absorbed to repair the structural deficit at the level of word structure.

The integration of contextual information into the process of creation and interpretation of a compound will be treated in chapter 3. The internal structure of the constituents of the compound has several grades of accessibility:

(a) Derivational accessibility. The verbal stem of a nominal constituent can be uncovered with its case frame. We assume that the access even to this structural centre is not trivial; this access involves "costs".

(b) Lexical accessibility. The internal semantic structure of a simplex can be retrieved. The question remains open, how such structures can be represented: by semantic features, semantic fields, networks, frames or stereotypes. We assume for our present purposes that the structure must allow for inference, i.e. the speaker and hearer should be able to infer more than just that which has been said, thus expanding and specifying their interpretation. Inferential procedures have been developed for semantic networks (cf. Wilks, 1980)[2] and also for stereotypes (cf. Eikmeyer and Rieser, 1981).

(c) Metaphorical expansion. If the solutions (a) and (b) do not work or if the context calls for a different interpretation, metaphorical processes can change the interpretation of the constituents, thus yielding another interpretation. I would be inclined to consider metaphorical expansion a special kind of attributional process (cf.

---

2 In Wilks (1980) *extraction* and extraction rules would correspond roughly to our concept of diffusion into a domain closed by a semi-permeable membrane.

Wildgen, 1982a).

Finally, every constituent of the compound (and possibly the compound as a whole) gives rise to associations (syntagmatic, paradigmatic, or owing to similarity of form or sound). These may constitute the material from which analogical processes (cf. chapter 4) borrow.

The activation of the procedures (a) to (c) can be influenced by contextual factors in such a way as to make a pair of coupled processes out of the two types of dissipative processes mentioned in thesis 2.

Part (a) of thesis 2 can be further specified.

Thesis 2*

The constituents of the compound can be structurally enriched if derivational, lexical and metaphorical structures are uncovered (in terms of diffusion, if this highly structured information is absorbed by the cell). On the basis of these enriched structures, that reading must be selected which fits best.

The complicated interaction between context, word semantics and the economy of language use cannot be dealt with here. Instead we shall turn to more specific features of nominal compounds in German and in related languages.

The most prominent structural feature of German (and English) compounds is their asymmetry. It is always the right-hand constituent (on the first level of segmentation) which determines the number, gender and (usually) the syntactic category of the compound. We can say that the right-hand constituent governs the grammatical features of the whole. This grammatical asymmetry has consequences for the semantics of compounds:

(a) The apparent symmetry of copulative compounds is unstable. Even with novel compounds the hearer tends to assign different meanings if the order of the constituents is inverted.

examples:

Mann-Frau ("man-woman")	≠	Frau-Mann ("woman-man")
Dichterkomponist ("poet-composer")	≠	Komponistendichter ("composer-poet")
Kardinal-Ökonom ("cardinal-economist")	≠	Ökonom-Kardinal ("economist-cardinal")

(b) In many compounds a tendency can be observed
- to place classificatory constituents at the right. These compounds approach derivations if these right-hand elements are reduced to a small list and are subject to morphological decay.
- to place evaluative constituents at the left, thus approaching prefixoids.

These synchronic fields of attracting forces can explain the diachronic fashioning of prefixes and suffixes out of constituents in nominal compounds. The poles of the compound exert a selective force which is dependent on the grammatical asymmetry of this construction.

These observations (cf. Wildgen, 1982b for more examples) lead to thesis 3.

Thesis 3

The grammatical asymmetry of the nominal compound in German constitutes a dynamic field with the poles *distinctive, evaluative* at the left and the poles *classificatory, general* at the right. This field leads synchronically to a typical distribution of constituents in the compound and diachronically to the creation of prefixes and suffixes out of constituents of the compound.

In thesis 2 and 2* we dealt with the case in which no relational element is realized in the compound and we said how the general law in thesis 1 is preserved under these circumstances. We may now consider the case in which two constituents of the compound are relational. As might be predicted, our analyses showed that three solutions to this problem are possible, and they depend on the internal structure of the constituents.

(a) The dynamic facts expressed in the relational terms (mostly verbs) are interpreted as parallel and simultaneous. As predicted by thesis 3, the more general term is placed at the right.

- SP/80//20/23/3
  Arbeits-Begräbnis ("working burial")
- SP/80/50/35/2
  Horch-Angriff ("listening-in attack")

These compounds are very rare.

(b) The second constituent (or its verbal centre) governs a subordinate clause; the verb of the clause is selected and appears as the left-hand constituent. This presupposes that the verb at the right can take subordinate clauses as complements.

- SZ/80/63/13/3
  Anzapf-Versuch ("attempt to tap")
- SZ/80/188/1/1
  Kontroll-Zwänge ("control restraint")

(c) In many cases the relational character of the left-hand element is not exploited and it is simply considered to be a nominal argument of the relation at the right. Thus the relational features of the right-hand element tend to be exploited before those of the left-hand one. This solution combines the effects of thesis 2 and 3.

Thesis 4

The asymmetry of the compound has the effect that a preference for the uncovering of lexical information in the right constituent exists.

To summarize, we can say that the local dynamics of compounds are very simple and can in fact be described in terms of structural diffusion and structural asymmetry. The fact that rather simple principles are at work could explain why nominal composition is learned before complex phrasal or sentential structures are acquired. Thus our dynamical treatment is more natural than the treatment within a classical transformational model.[3]

2. Models for local processes

It is not possible to review the long tradition of model-building for nominal composition here. As a general feature, we can say that the modelling of nominal composition has always been subsidiary to model-building in adjacent fields (either morphology or syntax). As a consequence of our argumentation in chapter 1 and in agreement with recent work in Artificial Intelligence (cf. Finin, 1980), we can state that local processes are either locally determined by the relational potential of their constituents and by the patterns summarized in the theses 1 to 4, or they are governed by global, i.e. textual and contextual, regularities and/or by analogical inferences. Thus the local aspect of nominal composition can be described on the basis of our arguments in chapter 1. There remains, however, a small class of noun-noun compounds which (a) can be interpreted out of context and (b) contain no traces of covert relational structures. In these cases the relational part has been eliminated and we must find a means of recovering it. Two possible ways of doing this can be proposed:

(a) The missing relational term can be inferred from our worldly knowledge. The constituents which are realized select a certain domain of knowledge, a knowledge frame, which allows us to infer one or more solutions to the problem of how these nominal constituents can be

3 It is, however, less technical. The technical appeal of transformational grammar still exerts a strong influence on many linguists.

connected.

(b) We possess an inventory of basic relational terms, i.e. relational atoms. The constituents make a selection from among these alternatives using certain affinities between nominal constituents and types of relations.

These solutions are not really contradictory. We assume that (a) is a broader position which uses a richer representation of culturally relevant knowledge. In (b) a smaller set of invariant frames is sought. As (b) is more economical and has a better chance of completion, we shall develop a new proposal for the construction of a universal list of basic (atomic) relational types.

In the studies of Brekle (1970/76), Kürschner (1974) and Levi (1978) the list of relational primitives was the outcome of (a) the choice of an a priori assumption and (b) a semi-systematic fitting to selected data.

The strategy we shall follow here is similar. Our a priori list is, however, based on recent results in the theory of dynamic systems. More specifically, we exploit the list of elementary and irreducible dynamic types classified by René Thom, Mather, Arnold and others. In short, our list can be derived in the framework of catastrophe theoretical semantics (compare Wildgen, 1982a). As we cannot develop this point of view formally and in detail (cf. Wildgen, 1979, 1981c, 1982a) we shall give only one illustrative example.

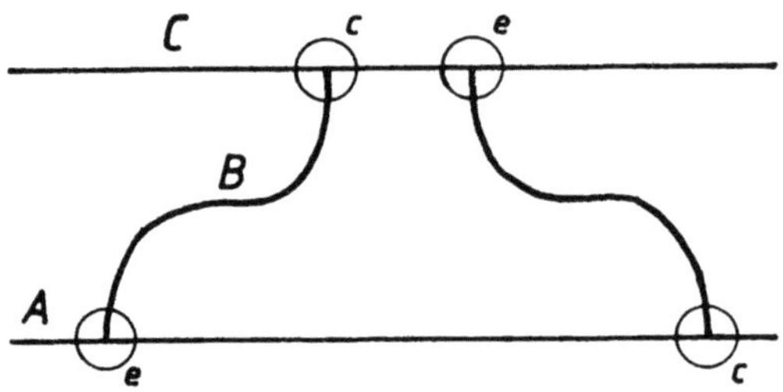

Fig. 1 Schema of dynamic interactions. A, B, C = attractors; e = catastrophe of emission, c = catastrophe of capture.

Fig. 1 shows a rather complex dynamical scheme, which can be derived from the elementary catastrophe called "butterfly". The heavy lines are interpreted as domains of stable existence of an entity. The bifurcations describe types of interactions such as:

e : *e*jecting, giving free etc.,

c : *c*atching, absorbing etc.

The whole "Gestalt" can be read as:

A gives B to C
C gives B to A

i.e. as *one* transfer cycle,
or instrumentally as
A affects C with his instrument B
(taking back the instrument).

Without going into the details of archetypal semantics, we can say that the following relations with two arguments (i.e. dynamic types with at least two stable attractors) can be derived

(a) Affecting	(A affects/influences/touches/.../B)
(b) Effecting	(A ejects/emits/creates/.../B)
(c) Instrumentality	(A affects C with the instrument B)
(d) Transfer	(A gives C to B / B receives C from A)
(e) Causation	(A causes C to affect B)
(f) localistic relations	(entering, leaving, being in, changing from A to B, changing from A to C via B)

Thus we can define a basic set of dynamic (= relational) primitives, whose features (structural stability, irreducibility) can be mathematically proved.

We can assume that Downing (1977) was right when she proposed regarding a similar list as the most probable candidates for the completion of the relational slot in the nominal compound and enriching it gradually, so that an open list was created. Within the framework of catastrophe theoretical semantics, we say that the fundamental attributional enrichment of the archetype is elaborated by sociocultural processes thus constituting the level of social and individual attributions (semantic prototypes and stereotypes; compare Wildgen 1982b and 1982d).

3. Global processes in nominal composition

The global processes involving the production and interpretation of nominal compounds can be seen as hierarchical strata, which on the one hand condense rather loose textual and sentential structures into a semantically rich and very short form. On the other hand these structures are not stable in the sense that they tend to disintegrate out of context. This loss of structure (loss of information, loss of negative entropy) can annihilate the whole process so that the compound is lost and forgotten. Under special circumstances, however, it can lead to a state of stability which does not conserve the initial richness of the nonce compound, but tends

(a) to select a basic but stable interpretation,
(b) to fit the result of contextual creation into the lexicon.

Thus the global processes exhibit two types of transitions (i.e. instabilities), which are qualitatively different.
Table 1 gives a schematic view of this idea.

stratum 1	discoursive organisation of content - the thematic structuring of texts and sentences (short term structuring)
	⇓ transition 1
stratum 2	condensation of content, operation of economic principles, stabilisation of contextual meaning (middle term structuring)
	⇓ transition 2
stratum 3	selection of structures at stratum 2 loss of unstable contextual keys, operation of economic principles in the knowledge system (long term structuring)

Table 1

Our empirical claims refer to the strata 1 and 2. As we have no data which allow hypotheses for the transition to stratum 3, we can only formulate some conjectures on this topic at the end of the chapter.

Our methodological strategies consisted in the establishment of a corpus of texts and transcribed dialogues such that the transition between stratum 1 and stratum 2 could be observed. It is clear that in many natural situations this transition is not made explicit and that shortcuts are chosen which lead directly from the global planning (silent language) to stratum 2 (or even to stratum 3, as in the creation of terminology).

The processes which were observed can be subdivided into two classes,

(a) Primary processes. These are largely independent of registers and specific types of texts and dialogues.

(b) Secondary processes. These mark specific textual and interactional structures and co-occur only with these types of texts and/or dialogues.

Table 1 gives the main classes of processes found in our empirical analyses:

Primary Processes		Secondary processes	
1.	Anaphorical processes,	1.1.	Alternation of variants,
1a.	anaphorical composition,	1.2.	descriptive elaboration,
1b.	cataphorical composition,	1.3.	enigmatic compounds,
1c.	partial anaphorical composition.	1.4.	marking of text-gestalts.
2.	Contrastive differentiation	2.1.	Contrastive intensification

Table 2

Two basic principles seem to govern these processes:

(a) The principle of variation. In the verbalisation of his

communicative intent (compare Wildgen, 1977: 45f) the speaker makes a choice from among a set of alternatives, weighting them along different scales. This range of alternatives can be further exploited when the speaker takes up the same thematic frame in the sequence of his utterances. We call this the principle of semantic-pragmatic variation. The set of alternatives is, however, not only exploited, it is also modified by every choice the speaker makes.

(b) The principle of framing. The speaker presupposes frames of interaction and of text-organisation (compare Goffman, 1976 and Kallmeyer & Schütze, 1977). The global processes of nominal composition do not only exploit such frames but also contribute to their constitution.

We shall now illustrate the different types of processes encountered (for a fuller treatment cf. Wildgen, 1981a).

(1a) The process of anaphorical composition.
The simplest case consists in the shortening of a nominal compound, by the elimination of a constituent (mostly in the center of the compound)

- Geldtransportauto ("money transport vehicle")
  ↓
  Geldauto ("money vehicle")
  SZ/79/257/13/1

- Sportbrieftauben ("sports carrier pigeon")
  ↓
  Sporttauben ("sports pigeon")
  SZ/80/91/1/1

More often the basic structure is not itself a compound but a phrase, a sentence or a text

- die Leute von Greenpeace ("The people from Greenpeace")
  ↓
  die Greenpeace-Leute ("the Greenpeace-people")
  SZ/80/54/1/1

	- Das Problem, wie man die *Gesichtszüge* eines heute lebenden Menschen auf seiner Photographie im Alter gewissermaßen *hochrechnen* könnte	("The problem of making an *extrapolation* from a person's photograph as to his *facial features* at a later date")
	▼ bei der Gesichtshochrechnung	("facial extrapolation")
	SZ/80/35/1/1	
(*a*)	- Bonns Innenministerium möchte *Kernkraftwerke* künftig *unter die Erde verlagern*	("The Ministry of the Interior in Bonn would like in future to put *atomic energy plants under the earth*")
(*b*)	▼ - Ein *Atommeiler unter der Erde*	("An atomic pile under the earth")
	▼ - ein *verbunkerter Reaktor*	("reactor in a bunker")
(*c*)	▼ - die *Atombunker*	("atomic bunker")
	▼ - ein *Untertage-Reaktor*	("underground reactor")
	▼ - die ... *Nuklearbunker*	("nuclear bunker")

The material analyzed in Wildgen (1981b) shows that the transformations from text, to sentence, to phrase and finally to the compound are both manifold and extremely complex. The basic principles are the variational resumption of a thematic complex and the reduction of the form by the elimination of noncentral elements in the utterance. The above example "Atombunker" showed how a set of novel compounds is created, all of which are candidates for a later lexicalization.

(1b) The process of cataphorical composition.

The processes known as cataphorical (cf. Herbermann, 1980) are in a sense merely variants of anaphorical processes; the difference is a pragmatic one. The speaker makes a jump to stratum 2 without verbalizing the previous stratum 1 (cf. Table 1). The hearer will either imagine adequate contexts using the situation and the local processes as inputs, or he will give up. In many cases the speaker feels obliged to help the hearer in his task.

We can distinguish two main types:

(*a*) The speaker/writer gives no help, the hearer must find the adequate interpretation for himself. We call these cases "enigmatic compounds".

- Schnittlauchfarm ("chives farm") — Münchner Abendzeitung, 18.10.80
- Dividenden-Arche ("dividend-arc") — (ibidem)

These compounds are stylistically marked and contribute to a kind of linguistic play which is typical for the marginal notes appearing in certain magazines.

(*b*) The speaker/writer helps the reader. This is done either immediately or later. In the second case the compound and its explication mark a textual gestalt (they constitute an act of gestalt-opening which is closed by the explication, compare Kallmeyer & Schütze, 1977, and Wildgen, 1981b).

Example

- Das Ein-Satz-Gutachten ∇ "Ihr müßt mehr arbeiten"	("The one-sentence verdict 'you must do more work'")
SZ/80/161/1/1	

- Im Zero-Nebel (title)	("In the Zero-fog")
∇ last sentence of the article ("Zero" is an art school)	
"Es scheint dringend geboten, den Zero-Nebel aufzulösen, der sich über die Szene gelegt hat"	("The dispersal of the Zero-fog, which has spread itself over the scene, would appear to be an urgent requirement")
SZ/80/28/12 (Feuilleton)	

(1c) Partially anaphoric composition.

Another variant of anaphorical processes occurs when only one of the constituents of the novel compound is connected thematically to the preceding text. This type can be used to preserve textual coherence, achieving, however, textual progression. This process is at the same time a starting point for analogical processes (cf. chapter four).

Example

SP/80/33/56f

Bildplatte	("picture record")
Bildplattenspieler	("picture record player")
Bildmaschine	("picture machine")
Bildplattengeschäft	("picture record business")
Bildplattenboom	("picture record boom")

All these processes are basically variational processes governed by pragmatic and even more general (thermodynamic) principles.

(2) The process of contrastive differentiation.
The second main class of processes, which we called "contrastive differentiation" is qualitatively different. Here an existing theme (or variational space in terms of the model developed in Wildgen, 1977) is divided into two (or, less often, three) subdomains which thus constitute a scale with opposed poles. In dynamic terms we can call this phenomenon a *bifurcation*. This process leads thermodynamically to an increase of structure, thus creating a state even farther from thermodynamic equilibrium than the initial state. The process of contrastive differentiation is a consequence of this structure-creating process. We shall give only a few examples.

Examples

SZ/80/99/1/1

Asyle des Landes ("the asylums of the country")

Kakao-	Tee-	Haferschleimheime
("cocoa-	tea-	gruel-homes")

SZ/80/65/1/1

die Welt ("the world")

Fortschrittszonen	Bewahrzonen
("progress zones")	("preservation zones")

SZ/80/70/1/1

die Laute der Henne ("sounds produced by the hen")

Konversationsgackern — Legegegacker

("conversational cackle") ("laying cackle")

SZ/80/91/1/1

Brieftauben ("carrier pigeons")

Sporttauben — Pharmatauben

("sports pigeons") ("pharmaceutical pigeons")

The secondary process of contrastive intensification is a special case, where first one pole of the split domain is elaborated in the text. The newly constituted bipolar (or tripolar) scale is now a context which influences the anaphorical process such that the original theme appears in a form which is adapted to the scale. Thus the simpler anaphorical mechanisms are deformed in the presence of a thematic bifurcation. The result of the bifurcation constitutes an attracting field with conflict.

Example
SZ/80/153/1/1

der liegende Auerochs (Bronzeplastik)
("the reclining aurochs (bronze statue)")

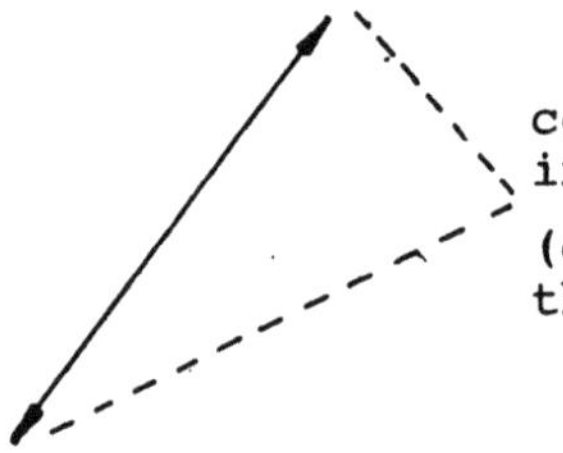

contrast: two living elks in the neighbourhood
(constitutes a context with the poles: living vs. dead nature)

das *Kunsttier*
("the artificial animal")

The process of contrastive differentiation exhibits a very basic dynamic principle, which can be encountered in many domains of language use (cf. Wildgen, 1981c).

Basic dynamic principle

Every unitary entity of form and meaning can bifurcate such that a scale with two poles (and eventually a metastable middle attractor) is created.

This principle can be formally derived from elementary catastrophe theory. The archetype of bipolar differentiation is derived from the cusp (and the archetype of trimodal differentiation is derived from the bufferfly) (cf. Wildgen, 1982a:55).

Thus the global process of nominal composition can be traced back to very deep biological regularities, in this way illustrating Thom's conjecture that language is the most highly developed and therefore purest natural morphology (in the biological sense, cf. Thom, 1974b).

The results obtained in this chapter lead to a preliminary hypothesis pertaining at the transition from stratum 2 to stratum 3.

Hypothesis 1

The anaphorical processes (1a, 1b, 1c), which basically exhibit the process of thermodynamic diffusion, will not normally lead to a stabilized compound on stratum 3. The secondary processes, however (not described here), can fill up the loss of information introducing connotational information and thus creating evaluatively rich structures. If the connotative contexts are stable or very suggestive, these compounds can achieve stability. This stability is, however, dependent on the stability of social evaluations.

Hypothesis 2

The process of contrastive differentiation as an innovative procedure is a possible source of changes in the lexicon of an individual or of a community. The result depends on the weight of the contextual need for differentiation and on the place the new compound is able to take in the system of lexical items.

4. Analogical processes in nominal composition

Analogical processes are global in the sense that, in the creation and interpretation of a compound, more structural information is used than that which is overtly contained in the compound and its constituents, or which can be inferred from them. Global processes are called analogical if this additional information is *linguistic* knowledge. The compound is *compared* to other linguistic forms or even to linguistic rules which have been extracted from linguistic experience. In a sense, this concept of analogical process is a challenge to generative grammar and its concept of *grammatical rule*. Rules are seen as the most general outcome of analogical processes. In fact it would be unwise to speak of analogical processes in nominal compounding if strict and systematic rules existed. But even Motsch (1979), one of the most prominent workers in generative word-formation in the sixties, acknowledges the fact that the rules of compounding are rather fuzzy and that they lack the generality, which one expects to be able to express in a grammar. He takes up the concept of "Analogiebildung" proposed by H. Paul in 1886. Paul states that rules are the most general form of analogical processes. If an analogy applies to many cases, a common characteristic can be abstracted; it becomes a rule (cf. Motsch, 1979: 30f). The range of repeated application of an analogy can vary: in the case of nominal composition very short ranges are to be found side by side with long ranges. On the basis of this very intuitive idea of Herman Paul's, we can establish a bipolar scale. At one pole of this scale, the compound is found to be identical (barring minor differences) to a compound which already exists in the lexicon of the hearer. We call this pole *"identity"*. At the other pole, the constituent structure of the compound selects a rule which predicts the result of the compositional process (the compound would be "Fregean" in the logical sense). We know that this scale is not balanced. Whereas the result "identity" is very common, the result "rule-governed" is very rare. As the experiments reported in Wildgen (1982c) show, most informants tend to overemphasize the pole "identity". Even novel compounds are classified as "known" if the hearer achieves a quick and simple interpretation. In spite of the fact that novel compounds are very frequent in German, they are generally not noticed to be new unless they are in some way stylistically marked.

As Geer and Gleitman (1972) have shown, many compounds are interpreted as wholes; thus linguistic knowledge of other forms is not activated in all instances. In Figure 2 we propose a scale which has only illustrative purposes. It shows how more and more linguistic knowledge is activated. The amount is related to the quantity of instances subsumed by an analogy and/or rule.

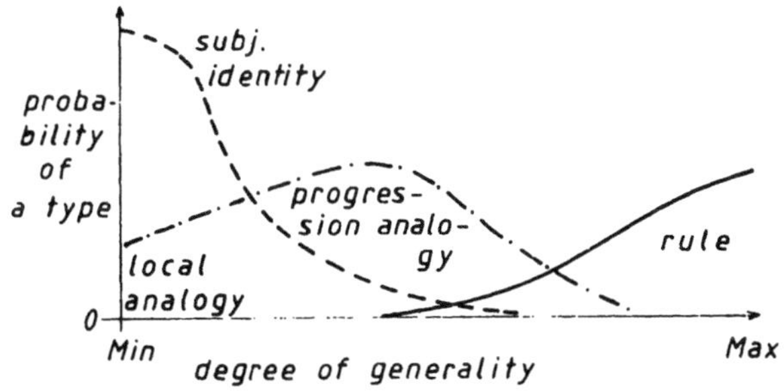

Fig. 2 Distribution of major types of nominal compounds

The curves in Fig. 2 are very roughly interpolated as no exact quantitative data are available. We have assumed that about 50 % of the compounds are known to the hearer/reader. The class of identity compounds is, however, very fuzzy; i.e. subjectively most of these compounds which can be interpreted easily are considered to be "known", even if analogical and regular processes have been activated. At the other end of the scale, regular compounds are only clearly differentiated from progression compounds if the analogical network becomes complicated and involves many instances. The probability of regular compounds depends also on the constituent structure of the compound. Thus the linearity of the scale in Fig. 2 is a simplification. In reality analogical processes involve many different aspects of the compound.

Local analogy and progressive analogy form *one* intermediate domain. The whole scale corresponds roughly to a trimodal field with intermediate, metastable attractor, which can be derived as an archetype from the catastrophe known as "butterfly" (cf. Wildgen, 1979). The intermediate domain is the most interesting one, as it seems to be the source for lexicalized *and* regular compounds. The extremes which are more stable have dominated the discussion in generative grammar where a bipolar methodological opposition appeared (in a sense as a dynamic consequence of Fig. 2). The poles were called "lexicalist -versus- transformationalist models of word-formation". In contrast to static generative grammars, our dynamic model-formation does not seek prestabilized fields; instabilities and transitions between stable systems reveal more of the principles underlying the whole phenomenon.[4]

The middle field of truly analogical processes can be further subdivided on the basis of our analysis of novel compounds in German.

(1) Local analogies

(a) Local analogy with a homophonous compound

- SZ/80/274/1/1

Preis-Bewußtsein	("prize-consciousness"; but the German word *Preis* has two meanings: "price" and "prize")

The choice of the reading 'prize' destroys the identity and creates a new compound. It is further assisted by other compounds (Preisträger, Kleist-Preis, Literaturpreis, Tausend-DM-Preis (prize-holder, Kleist-prize, literature-prize, thousand-mark-prize)) in the text.

Other examples are:

Auto-Bahn	("railway (Bahn) for transporting cars (Auto)"; but the word *Autobahn* usually means

4 In a sense the modelling of language in Thom's perspective constitutes a new paradigm, whereas generative grammars, logical grammars, AI-systems are variants of structuralism (cf. Givon, 1979: 1-43).

	"motorway" (cf. Holst, 1980: 8)
Hosen-Träger	("someone who wears (tragen) trousers (Hose)"; but *Hosenträger* usually means "braces")

SZ/80/148/17

(b) Local analogy by expansion

Brotgeber	("giver of bread")
Brot-für-die-Welt-Geber	("giver of bread-for-the-world")
- SZ/80/271/1/1	
Nulltarif	("free fares")
Null-Benzintarif	("free petrol fares")
- SZ/80/40/1/1	

(c) Local analogy by partial variation

- SZ/80/68/17/1	
Frühaufsteher	("early riser")
Früh-Frühstücker	("early breakfaster")
- SZ/80/52/1/1	
Fehlerquoten in *Flach-* sowie *Hoch*rechnung	("rate of error in *super-* and *extrapolation*")
- SZ/80/118/1/1	
Alleinjogging	("solo jog")
Alleingang	("solo run")
- SZ/80/250/1/1	
Lautverstärker	("loud-amplifier")
Lautsprecher	("loud-speaker")
Verstärker	("amplifier")

In general, partial analogies tend to change the first constituent rather than the second. This seems to be a consequence of thesis 3 in chapter 1.

(d) Local analogy referring to phraseologisms

As phraseologisms correspond functionally to lexicalized

items, it is natural that analogical processes operate on them. The technique usually corresponds to the expansion strategy (b).

- SZ/80/265/1/1

  Eine Stimme aus der *Umweltwüste* ... mahnt ("a voice in the *desert environment* ... admonishes")

- SZ/80/210/1/1

  Auf die goldene *Fairneßwaage* legen ("to lay upon the golden *scales of fairness*")

- SZ/80/57/1/1

  Die *Resolutionstrommel* rühren ("to beat the *resolution drum*")

- SZ/80/33/1/1

  Den *Schwimmgürtel* enger schnallen ("to tighten one's life-belt")

The last compound *Schwimmgürtel* changes the original concrete reading by the fact that '*Schwimm*' modifies the whole phraseologism. The whole phrase means that the costs for the upkeep of a swimming-bath are to be reduced. Thus the expansion of the noun in the phraseologism achieves a very general modification of the phrase instead of modifying only the noun.

(2) Progression analogy

The problem of progression analogy lies in the fact that every step in the progression may change the analogical basis. Thus, looking only at the surface forms of compounds, we may recognize the recurrent use of one constituent in one position. However, as the other position is changed, different microprocesses operate and different semantic and/or stylistic structures appear. Thus the basis of analogy is easily destabilized. This explains the fact that larger series of progressive analogies tend to become more and more heterogenous, such that the unity of the series is destroyed. In addition, we must take into account that not every repetition of a constituent in another compound constitutes a progression, as this recurrence can be independent of the first occurrence.

There are, however, some situations in which the existence of a progression can be clearly proved. We can distinguish three types:

(a) The progression set starts from a small set of analogical compounds and expands very quickly. The weight of this progression is such that the members of a speech community can recognize the internal coherence of the set. In German we can diagnose many prominent progression series.

- Muffel ("someone who dislikes a particular thing"); originally: *Krawattenmuffel* ("someone who dislikes ties") in the fifties;
- Krise ("crisis"); *Ölkrise* ("oil crisis") as a centre.

Krisen-	("crisis-")
Atom-	("atomic-")
Polit-	("political-")
Öko-	("eco-")

The constituents *Polit-* and *Öko-* are already approaching the status of prefix. In general we can say that large progression series can change the status of a constituent, making it a prefix or a suffix (along the dynamic field mentioned in thesis 3).

(b) If the progression series is less prominent it may be caused simply by the prominence of a certain theme in specific texts (at certain periods). A special case exists if a set of analogical compounds appears in one rather restricted context. Such a set can directly appear as a list.

Example

(i) A series of books on photographical techniques is advertized by a list:

Makrofotographie	("macro photography")
Sport-	("sport photography")

Tabletop-	("tabletop photography")
Reise-	("travel photography")
Kinder-	("child photography")
Blitz-	("flash photography")
Mädchen-	("photography of girls")
Tier-	("animal photography")

This series can even be regarded as iconic: the progression set gives a picture of the series of books edited by a firm. The semantic structures are, however, not identical in this list, i.e. *Kinder, Mädchen, Tier* are objects of the verb embedded in the second constituent whereas *Sport, Reise* are domains in which motifs are sought, *Makro* refers to a photographical technique and *Tabletop* to a method of constructing imaginative landscapes. Thus the analogy is primarily an analogy of linguistic form.

(ii) In Wildgen (1980b) we analyzed about 200 nominal compounds having *Strauß, Schmidt* and *Kanzler* as constituents. These compounds were found in 74 reports of the German magazine "Spiegel", in the period 5.11.79 to 8.7.80. The magazine reported on the controversies between *Schmidt* and *Strauß*, who were candidates for the German elections in autumn 1980. In this case a special occasion (the elections) and a special type of text (reports on the election campaign) constituted the pragmatic background for a large progression series. The coherence of the progression is, however, less clear than in case (i). The coherence of a series is a measure dependent on the coherence of the situation and on the copresence of the elements in a restricted context. In our corpus of novel compounds we have found 72 constituents which are used to form series of at least three elements. The most frequent constituents are listed below.

-Krieg- ("war")	N = 16	Auto-Krieg ("car-war") Kriegswolke ("warcloud")

- -

-Atom- ("atomic")	N = 14	Atomalarm ("atomic alarm") Atomziel ("atomic goal")
-Computer- ("computer")	N = 14	Computer-Pionier ("computer-pioneer") Kopf-Computer ("head computer")
-Zeit- ("time")	N = 14	Zeit-Salat ("time salad") Hamster-Zeit ("hamster time")
-Boykott- ("boycott")	N = 13	Boykott-Einfall ("boycott-brain-wave") Essens-Boykott ("meal-boycott")

(iii) A third group of nominal compounds uses strongly relational nouns as one constituent. As these nouns are used very frequently to form a compound, a progression series is produced. This case is on the borderline with regular compounds, as the series can be explained by the facts that:

(a) a systematic microprocess is applied,
(b) the set of lexical elements fitting this process is small.

Examples
In our corpus we found the following relational constituents ordered by the frequency of their occurrence in novel compounds.

Mutter ("mother")	-Mutter (Mütter)	10	Heldenmutter ("mother of a/the hero")
	Mutter-	22	Muttergift ("mother poison")

Vater ("father")	-Vater	7	Zoovater ("zoo-father")
	Vater-	3	Vater-Krise ("father-crisis")
Tochter ("daughter")	-Tochter	8	Banktochter ("daughter-bank")
	Tochter-	1	Tochterbombe ("daughter-bomb")
Sohn ("son")	-Sohn	4	Carter-Sohn ("Carter-son")
Bruder ("brother")	-Bruder (Brüder)	5	Boykott-Bruder ("boycott brother")
	Bruder-	4	Bruder-Wohnung ("brother's flat")
		64	

Freund ("friend")	-Freund	15	Britenfreund ("Britain-friend")
Feind ("enemy")	-Feind(in)	7	Busenfeindin ("bosom enemy")
Gegner ("opponent")	-Gegner	8	Asylgegner ("asylum opponent")
Fan ("fan")	-Fan	12	Nudelfan ("noodle-fan")
	Fan-	3	Fan-Kurve ("fan-curve")
		45	

Chef ("boss")	-Chef	7	BP-Chef ("BP-boss")
	Chef-	6	Chef-Puritaner ("head Puritan")
Kollege ("colleague")	-Kollege	4	Staats-Kollege ("state-colleague")
	Kollege(n)-	2	Kollegen-Schelte ("the scolding of a colleague")
		19	

Table 2

As table 2 shows, the list of very productive constituents is rather small. They all belong to the domain of social relations. The first group contains terms denoting immediate kinship relations. The second group subsumes positive and negative social relations and the third group contains terms for relations in the domain of professions.

As our analysis shows, these forms of social relationship cannot simply be subsumed under the general label "relational". They constitute a special subdomain of the lexicon. Only a few of the compounds are based on microprocesses exploiting the valencies of these nouns. Many of these compounds are instead based on analogy to existing lexicalized forms and they can therefore fill the open position with arguments other than immediate arguments of the relation.

Examples

Einser-Töchter	("daughters who always received grade one (Einser)")
Mütter-Töchter	("mother-daughters")
Auslandstöchter	("daughters abroad")
Kindergartentöchter	("daughters in the kindergarten")

Only the constituents *Fan, Gegner , Freund, Feind* are systematically relational. These observations falsify some of the claims in Fanselow (1981). Relational structures are only inherited by the compound under special circumstances. It

seems that the discovery and explanation of many details of the local processes depends on an extensive analysis of global processes. Thus the results of this and the foregoing chapters may influence further elaborations of the principles found in the first chapters of this article.
The results of this chapter are summarised in thesis 5.

Thesis 5

The dynamic field underlying the use of nominal compounds can be modelled by a trimodal field (identity-analogy-rule). The metastable domain, which is the site of analogical processes, is a source for both lexicalization *and* semi-regular patterns of composition.

5. Diachronic perspectives

In a dynamic model, the division established between synchrony and diachrony by structuralist methodologies will naturally be abolished. Our empirical basis contains, however, only a very narrow historical dimension, such that we can only formulate some assumptions on the diachronic dynamics of nominal composition.

In thesis 3 and thesis 5 we have already mentioned fundamental dynamic mechanisms which can be exploited in language change. We assume that many of the important diachronic changes can be explained by these mechanisms.

If we look at other languages, we remark a fundamental inequality. Some languages such as German, English, Chinese and many others make extensive use of compositional processes, whereas others have only very poor compositional patterns. This typological difference is, however, rather unstable, i.e. languages can change the amount of composition accepted as normal and many language families are heterogeneous in this respect. A rather peculiar effect has been reported from Australian languages. Whereas the Western Australian languages prefer to create new roots for the verbal lexicon, thus developing a rather extensive inventory of lexematic units, the Eastern languages prefer to combine elements of a smaller set of verbal roots to form verbal compounds. One could put forward the hypothesis that Australia, as a territory which was not "perturbed" by immigrations and population drifts for millenias, has segregated two fundamental alternatives for the assembling of lexical units, alternatives which normally are not stable enough to characterize a farreaching line of subsequent grammatical systems.

If a speculative outlook is allowed, we could parallelize the three fundamental types of lexical innovation with fundamental biological processes (compare Kaplan, 1978).

lexical innovation	genetic innovation
(a) creation of new roots	(a) genetic mutations
(b) composition	(b) recombination of "successful" mutands
(c) lexical borrowing (language mixture)	(c) heterosexual exchange of genetic material

Table 3

In the Australian example, the two fundamental mechanisms could develop and gain wider expansion (by diffusion) by the fact that lexical borrowing was restricted. In our western languages, where a stable dominance of (a) or (b) could not develop, all existing languages are the result of extensive borrowing and polycentric diffusion of structures.

These final perspectives show that a strict dynamical modelling opens a vast domain of interesting perspectives, where new solutions can be found.

A Note on the Regensburger Corpus of New Nominal Compounds

This article is an analysis of data which I have gathered in the context of the DFG-research project "Aktualgenese nominaler Komposita" in Regensburg (headed by Prof. H. E. Brekle). The final corpus can be purchased on microfilm.

The abbreviations behind the examples cited in this study refer to the source of the nominal compound. They come mainly from articles published in the journals:

Süddeutsche Zeitung (SZ) and Spiegel (SP);
the numbers refer to the year of publication, the no. of the issue, the page and/or column where the compound can be found.

Example:
SP/80/33/56 = Spiegel / 1980 / issue no. 33 / page 56
SZ/80/65/1/1= Süddeutsche Zeitung / 1980 / issue no. 65 / page 1 / column 1

References

Arnold, V. I.: 1972, "Normal Forms for Functions near Degenerate Critical Points, the Weyl Groups of A_K, D_K and E_K, and Lagrangian Singularities". In: *Functional Analysis and its Applications*, 6, pp 254-272.

Bloom, L./Lahey, M.: 1978, *Language Development and Language Disorders*. New York.

Brekle, H. E.: 1970/76, *Generative Satzsemantik im System der englischen Nominalkomposition* (second edition 1976). Fink: München.

Dixon, R. M. W.: 1980, *The Languages of Australia*. University Press: Cambridge.

Downing, P.: 1977, "On the Creation and Use of English Compound Nouns". In: *Language*, 53, pp. 810-842.

Dressler, W.: 1977, "Elements of a Polycentristic Theory of Word Formation". In: *Wiener Linguistische Gazette*, 15, pp. 13-32.

Eikmeyer, H.-J./Rieser, H.: 1981, "Meanings, Intensions, and Stereotypes. A New Approach to Linguistic Semantics". In: Eikmeyer/Rieser (eds.), *Words, Worlds and Contexts*. New Approaches to Word Semantics. De Gruyter: Berlin, pp. 133-150.

Fanselow, G.: 1981, "Neues von der Kompositafront oder zu drei Paradigmata in der Kompositagrammatik". In: *Studium Linguistik*, 11, pp. 43-57.

Finin, T. W.: 1980, *The Semantic Interpretation of Compound Nominals* (Report T-96, Coordinated Science Laboratory, Urbana). University of Illinois.

Fleischer, W.: 1969, *Wortbildung der deutschen Gegenwartssprache*. VEB Bibliographisches Institut: Leipzig.

Fleischer, W.: 1979, "Kommunikativ-pragmatische Aspekte der Wortbildung". In: I. Rosengren (ed.), *Sprache und Pragmatik* (Lunder Symposion 1978). Lund, pp. 317-329.

Geer, S.E./Gleitman, H.: 1972, "Paraphrasing and Remembering Compound Words". In: *Journal of Verbal Learning and Verbal Behavior*, 11, pp. 348-355.

Gilmore, R.: 1980, *Catastrophe Theory for Scientists and Engineers*. Wiley: New York.

Givon, T.: 1979, *On Understanding Grammar*. Academic Press: New York.

Gleitman, L.R./Gleitman, H.: 1970, *Phrase and Paraphrase.* Some Innovative Uses of Language. Norton: New York.

Goffman, E.: 1976, "Replies and Responses". In: *Language and Society,* 5, pp. 257-313.

Herbermann, C.P.: 1980, *Wort, Basis, Lexem und ihre Rolle bei der Bildung komplexer Substantive.* Fink: München.

Holst, F.: 1980, *Okkasionelle Umdeutungen von sprachlichen Ausdrücken* (Ms). Köln.

Kallmeyer, W./Schütze, F.: 1977, "Zur Konstitution von Kommunikationsschemata der Sachverhaltsdarstellung". In: B. Wegner (ed.), *Gesprächsanalysen.* Buske: Hamburg, pp. 159-274.

Kaplan, R. W.: 1978, *Der Ursprung des Lebens. Biogenetik, ein Forschungsgebiet heutiger Naturwissenschaft.* Thieme: Stuttgart

Kürschner, W.: 1974, *Zur syntaktischen Beschreibung deutscher Nominalkomposita.* Niemeyer: Tübingen.

Levi, J. N.: 1978, *The Syntax and Semantics of Complex Nominals.* Academic Press: London.

McCune-Nicolich, L.: 1980, "The Cognitive Basis of Relational Words in the Single Word Period" In: *Journal of Child Language,*8, pp. 15-34.

Motsch, W.: 1979, "Zum Status von Wortbildungsregularitäten". In: *DRLAV* (Documentation et recherche en linguistique allemande contemporaine-Vincennes), 20, pp. 41-67.

Poston, T./Stewart, J.N.: 1978, *Catastrophe Theory and its Applications.*Pitman: London.

Okubo, A.: 1980, *Diffusion and Ecological Problems: Mathematical Models.* Berlin

Rubinow, S. I.: 1980, *Introduction to Mathematical Biology.* Wiley: New York.

Thom, R.: 1972/77, *Stabilité Structurelle et Morphogénèse. Essai d'une théorie générale des modéles.* Benjamin: New York (translation 1975, Wesley: New York; second edition, augmented and corrected 1977, Interéditions: Paris).

Thom, R.; 1974a, *Modéles mathématiques de la morphogénèse.* Recueil de textes sur la théorie des catastrophes et ses applications. Union d'éditions, series 10-18: Paris.

Thom, R.: 1974b, "La linguistique, discipline morphologique exemplaire". In: *Critique,* 30 (1), pp. 235-245.

Wildgen, W.: 1977, *Differentielle Linguistik, Entwurf eines Modells zur Beschreibung und Messung semantischer und pragmatischer Variation.* Niemeyer: Tübingen.

Wildgen, W.: 1979, *Verständigungsdynamik: Bausteine für ein dynamisches Sprachmodell* (Habilitationsschrift). Regensburg.

Wildgen, W.: 1980, "Textuelle Bedingungen der Einführung und Verwendung ad hoc gebildeter Komposita". In: *L.A.U.T. Preprints*, Series A, Paper No. 80.

Wildgen, W.: 1981a, "Grundstrukturen und Var tionsmöglichkeiten bei Eigennamenkomposita: Komposita mit den Eigennamen *Schmidt* und *Strauß* als Konstituenten in Wahlkampfberichten des *Spiegels*". In: *L.A.U.T. Preprints*, Series A, Paper No. 82, Trier.

Wildgen, W.: 1981b, "Makroprozesse bei der Verwendung nominaler *ad hoc*-Komposita im Deutschen". In: *L.A.U.T. Preprints*, Series A, Paper No. 83, Trier (to appear 1982 in: *Die Sprache)*.

Wildgen, W.: 1981c, "Archetypal Dynamics in Word Semantics. An Application of Catastrophe Theory". In: H. Eikmeyer/ H. Rieser (eds.), *Words, Worlds and Contexts*. De Gruyter: Berlin, pp. 234-296.

Wildgen, W.: 1981d, "Semantic Description in the Framework of Catastrophe Theory". In: B. Rieger (ed.), *Empirical Semantics*, Vol. II. Brockmeyer: Bochum, pp. 792-818.

Wildgen, W.: 1981e, "Modelling Vagueness in Catastrophe Theoretical Semantics". In: T. Ballmer/M. Pinkal (eds.), *Approaching Vagueness*. North Holland: Amsterdam (to appear 1982).

Wildgen, W.: 1982a, *Catastrophe Theoretical Semantics. An Elaboration and Application of René Thom's Theory*, Series: Pragmatics and Beyond: III, 5). Benjamins: Amsterdam.

Wildgen, W.: 1982b, "Zur Dynamik lokaler Kompositionsprozesse: Am Beispiel nominaler *ad hoc*-Komposita im Deutschen". In: *Folia Linguistica*, XII.

Wildgen, W.: 1982c, *Psycholinguistische Experimente zur Verarbeitung und subjektiven Einschätzung nominaler Komposita*, Report given at the 3rd Colloquium of Semiotics in Hamburg (1981).

Wildgen, W.: 1982d, *Skizze einer katastrophentheoretisch fundierten dynamischen Semantik*, Report given at the 4th Congress of the DGfS, Cologne 1982.

Wildgen, W.: 1985. *Archetypensemantik. Grundlagen für eine dynamische Semantik auf der Basis der Katastrophentheorie*, Narr, Tübingen

Wilks, Y.: 1980, "Frames, Semantics and Novelty". In: D. Metzing (ed.), *Frame Conceptions and Text Understanding*. De Gruyter: Berlin, pp. 134-163.

Woodcock, A./Davis, Am.: 1978, *Catastrophe Theory*. Dutton: New York.

PART TWO: THE EMPIRICAL STUDY OF PROCESSUAL ASPECTS OF LANGUAGE

DYNAMIC SEMANTICS

Walther Kindt
Universität Bielefeld

Abstract

The first proposals for a modelling of the dynamic aspects of interpretations were put forward in, for example, Ballmer 1973 and Kindt 1974. So far, however, such initiatives have scarcely been taken up in linguistic semantics. That the majority of semanticists persist in working with static models does of course have reasons. Chief among these are the persistent influence of common place theories about semantics, insufficient striving towards empiricism, the fact that empirical access to the internal processes of interpretation in participants in communication can only be gained indirectly, and the fact that interactive parameters of interpretations have not been taken into account adequately. Beside giving more precise reasons for this assessment of semantics up to the present I should like in this article to sketch out a general yet also simple framework in which the dynamics of interpretation processes can be dealt with both theoretically and empirically; at the same time the special role of the interactive aspect of meaning constitution will be enlarged upon.

1. Why do we need a Dynamic Semantics?

For several years now in the Federal Republic there has been public discussion of a central problem of semantics without linguists or semanticists taking the expected part in it. I refer to the problem of comprehensibility, which plays an important role in social politics, for example in the communication between citizens and government: government forms are frequently so incomprehensible to citizens that relations between the groups concerned are subject to considerable strain.

On the government side attempts are being made to overcome these difficulties; for example books and brochures are published containing recommendations on how to achieve comprehensibility of texts. The remedies proposed in them are, however, based only to a low degree on the more recent findings of linguistic semantics, in the main they stem from dogmas of traditional grammar or the results of psychological research. I attribute this state of affairs paramountly to the lack of interest and involvement of semanticists in practical problems and also their inability to counter the partially deficient formulas of the practicians with more suitable methods. This inability is connected in turn with the fact that, just as in commonplace theories of language, in linguistics too a static view of language is still largely predominant, and in particular the dynamic character of meaning constitution goes unrecognized. The inadequacy of a static view of semantics can be clearly seen for example in much of the common advice about language, which instead of increasing comprehensibility restricts creative scope in lanuage use in an unacceptable manner.

The efforts referred to above at improving administrative language are not a new development. Thus for example the first edition of the handbook "Fingerzeige für die Gesetzes- und Amtssprache" (Hints on Legal and Official Language) (cf. Daum 1980), a book widely known in the Federal Republic and recommended personally by the Minister for the Interior, dates in fact from the year 1930. This probably also explains why certain normative semantic positions in traditional cultivation of language have continued to have an effect up to the present day. If one ignores for a moment their negative influence on the concrete linguistic behaviour of speakers one can again be amused at certain contradictions between the linguistic recommendations of the guardians of language and their own linguistic behaviour. On the one hand, namely, certain linguistic formulations are sometimes proscribed as being logically inadequate, because there is said to be incompatibility of the interpretations of different parts of the formulation. Thus, for example, in the above-mentioned handbook the phrase "den Stand-punkt vertreten, daß ..." (support the viewpoint that ...) is criticized on the grounds that: "Einen Punkt kann man strenggenommen weder vertreten noch vorbringen. Auch ist es unlogisch, an ihn einen daß-Satz anzuschließen." (one cannot strictly speaking either support or advance a point. Additionally it is illogical to attach a that-clause to it) (cf. Daum 1980: p. 11). On the other hand the language points themselves constantly employ expressions which could be criticized in a similar manner; thus, for instance, one ought not to say that "Sprache gedankenlos dahin plätschert" (language babbles thoughtlessly along) (cf. Daum 1980: p. 13), for language, as we know, is neither a liquid substance nor is it capable of thought.

Judged as a whole, a large part of the linguistic recommendations in Daum 1980 is unacceptable as they assume a much too inflexible handling of standard interpretations of linguistic expressions. There are, it is true, boundaries, independant of time and group specific, to the tolerance of unaccustomed expressions, the flexibility of the present meaning constitution and the speed with which new interpretations are adapted are, however, much greater than is generally assumed. At the same time it cannot be the job of linguistic recommendations to damn certain expressions of language absolutely as impeding understanding or as stylistically inappropriate; rather it can at most be a question of making it clear that comprehensibility is to be viewed closely connected to the establishment of context and that consequently deviation from the standard is permissible if suitable contextual conditions are provided for. We can see exemplarically from the points mentioned here of flexibility and context-independence of meaning constitution that the real problems of comprehensibility can only be dealt with adequately within the framework of a dynamic conception of semantics.

We should not be surprised that the ordinary views of semantics are largely of a static nature; for it would not be helpful to necessary processes of understanding if the participants in communication were aware of the real extent of the flexibility and complexity of interpretative processes. But why have linguists not so far freed themselves of such static views and applied themselves to the development of dynamic conceptions of semantics? The chief reason for this seems to lie in the semanticists' inadequate knowledge of empiricism. The procedure chiefly practised so far of discussing examples with recourse to one's own linguistic intuition is quite inadequate for further development and testing of models in semantics, because the contextually regulated profuseness and variability of interpretations cannot be consciously controlled even by the most sensitized semanticist. Bringing it down to a short formula: without stronger empiricism of semantics it will not be possible to develop dynamic models; without dynamic modellings, however, no effective semantic theories applicable to practical problems will be obtained.

2. Prerequisites for an Empiricization of Semantics

Independently of the view that the development of a dynamic semantics is necessary, efforts in semantics in my opinion have now reached a point where greater advances are scarcely possible without systematic empirical research (cf. Kindt 1979). If this thesis is correct we must ask under what conditions and in what theoretical framework empirical semantics can be carried on at all. In my judgement, which I cannot substantiate here in detail, the semantic conceptions established so far are little suited as the starting point for an empiricization. There are three aspects I would like to call attention to here.

In semantics it is usual to work with meanings as special abstractions. The exact ontological status of these abstractions, however, remains to a large extent unclear in both a theoretical and an empirical respect. On the one hand it should be asked what kind of objects meanings actually are. On the other hand it should be made explicit on the basis of what empirical relations the abstractions concerned have come into being. This explicitness requirement is not fulfilled at all by the present conceptions of semantics. I am, certainly, of the opinion that some of these abstractions can be empirically legitimized in principle; as long as such a legitimation is not available in a formulated form, however, I believe we should fall back in the present situation on the position that the only valid objects of analysis in semantics are the 'real' meanings which are actually established by communication participants in particular situations (cf. Kindt 1981). I particularly oppose the frequently practised sub-division into literal and pragmatic meanings; this in turn not because I would deny that the division has a certain core of truth but because I believe it leads to an immunization of semantic research against empirical demands. At the same time I consider this division to be inappropriate for research strategy because real interpretation processes with larger linguistic units do not work in such a way that the literal meaning is ascertained first and subsequently additional, pragmatically determined interpretation processes are joined on.

Requiring that semantics be linked back to the investigation of real meanings remains itself, however, 'grey theory' unless it is coupled with statements as to how we can empirically measure whether a communication participant has assigned a particular meaning to an expression or not. Semanticists have yet to become aware of this measurement problem and accordingly the prevailing measuring procedures, especially from psycholinguistics and conversation analysis must be taken up and systematically developed further.

In accordance with my remarks in section 1. semantics finally needs empirically serviceable theoretical models to replace the present, much too static approaches to meaning constitution. Such models must on the one hand be well-founded sociologically: in this the social restrictedness of meanings is to be related not only to the acquisition of meaning during socialisation and the dependence on socially-defined situational factors, but meaning constitution must itself be regarded as a central social activity, which is, admittedly, dealt with in a very flexible manner but is at the same time kept under constant, interactive control by procedures practised during socialisation. On the other hand such models must reflect the decision processes, which remain largely implicit in communication, in the choice of reading and the construction of 'new' meanings making full use of the available scope for vagueness (cf. Kindt 1982a,b).

The three aspects emphasized here should make it clear how far we are from the establishment of an empirical semantics. As regards the aspects of model development, however, I would like to give a more exact assessment of the situation in the following.

3. Modelling Tasks of Dynamic Semantics

3.1 If one wants to employ the distinction dynamic vs. static semantics as more than just an eye-catching concept for a characterization of semantic models, then the degree of dynamics of a model should, it seems, be measured according to how explicitly the processes of meaning construction are really reflected in it and to what extent the flexibility of meaning constitution is covered over and above the consideration of standard meanings in standard contexts. In this sense the current semantic models are to be termed static for the following reasons. On the one hand they only apply contextual factors in a very global way and do not reconstruct their influence on interpretations explicitly or else they shift the actual job of modelling on to pragmatics. On the other hand they start from far too restricted meaning inventories for linguistic expressions and syntactic constructions, or they scarcely notice the procedures of interpretational modification constantly employed by communication participants. If, by contrast, it is the job of dynamic semantics to describe explicitly the constructional procedures of interpretations and to account for the flexibility of meaning constitution, the first step must be to investigate systematically what interpretational procedures are in fact used by communication participants and what communicational conditions make the flexibility of meaning constitution possible. But on the other hand to be able to explain too how communication participants reach an understanding, i.e.

a common area in meaning constitution, the second step must be to investigate according to what rules the interpretation procedures are put into action and in what way the flexibility assumed can again be restricted. Both investigative steps must, moreover, be differentiated as to the treatment of questions referring on the one hand to the internal level of individual language processing and on the other to the external level of interactive control of interpretation processes. More concretely this means for the first step in the investigation that on the one hand the variety of the fundamentally applicable meaning construction procedures which are available to each and every communication participant is to be wrought out; on the other hand the interactive procedures are to be described which operate above the individual results of understanding and make possible interpretations going beyond them. Similarly too for the second investigate step a distinction must be made between questions concerning the internal and the external level. On the internal level it is asked how similarity of context conditions and of rules for the use of interpretation procedures affects a common area in the meaning constitution of different participants; for the external level it must be made clear according to what interactive procedures initially differing, individual interpretations can be adapted to one another.

3.2 The tasks just described seem to make the development of a very complex interpretation model necessary. Following a strategy often employed by linguists one could set about naming different sub-components of the model, characterizing their status vividly and establishing as many differentiations as possible needed to explain the whole of the underlying phenomena. In my opinion an empirically and theoretically effective strategy aims in precisely the opposite direction: certainly one must keep the differentiatedness of the phenomena in mind, but in general a simple and successful modelling only comes about by idealising and generalising. More precisely it cannot be a question of the development of a single model satisfying at once all differentiation requirements; rather it is expedient to plan model building as a process of designing models and successively refining them, while retaining the applicability of the coarse models as well and when a case occurs for application selecting the coarsest and simplest model which can supply the information required. To this extent there is no conflict with what was said before if I here first of all press for the formulation of a model conception which is as simple as possible.

That communication participants' assignments of meaning are context-dependent is a triviality. To take this dependency into account long lists of context factors have been made in linguistics and for interpretation functions a corresponding number of arguments have been formally introduced. The purely schematic differentiation of the context parameter into 8 or, if you like, 27 individual factors is, however, of no use for the modelling of interpretation processes: whether the interpretation function provided for here is planned as a function over one or several context variables is of no importance as long as additional data are not given at the same time about the influence of the different variables on the behaviour of the function and thus no interesting properties of this function can be logically deduced. In this sense the formal introduction of several context factors simulates a richness in the semantic model which is not met from an empirical point of view. On a closer look the concept of context proves in any case to be one of the particularly unreflected and non-uniformly employed termini of linguistics. As a basis for an explanation of this term a concept of situation is required. Then contexts can be introduced as abstractions from situations, i.e. formulated set-theoretically as classes of situations (cf. Kindt 1982 a, b). However, it remains unclear which situational conditions should be abstracted from in the transition to contexts and which conditions are to be regarded as interpretationally relevant. For example the lighting conditions under which a text is read certainly represent a situational factor which has a decisive influence on text reception; at the same time, however, one would not want to admit this factor unconditionally as a context factor. I do not want to go into more detail about this problem of demarcation, as the introduction of an explicit concept of situation is sifficient for the following presentation. In principle model theory together with the concept of structure (cf. Ebbinghaus et al. 1978) provides a suitable approach to this. However, this concept of structure is too weak for the purposes of linguistic semantics for the following reasons. Firstly, in this concept changes in the properties or classification of objects cannot be distinguished from changes in language interpretation; for example it makes an important difference for the semantic analysis of the sentence uttered at a point of time t "Herr Schmidt ist ein Bauer" (Mr. Schmidt is a peasant) whether Mr. Schmidt changed his job at some point shortly before t and became a peasant or whether he is called a peasant at the point of time t because of his indelicate behaviour. Secondly this concept is static because only one interpretation is admitted for each linguistic expression within a structure. In this way, however, treatment is excluded of sentences like e.g. "Kohl ißt und redet gern Kohl" (Kohl likes to eat cabbage and talks nonsense) where "Kohl" is used in three different meanings (Kohl = the leader of the CDU, Kohl = the vegetable, cabbage, Kohl = nonsense). Thirdly the model-theoretical concept of structure does not take into account the central semantic property of natural languages, that the interpretations of different linguistic expressions have a mutual influence on one another. In the sentence "Kohl schmeckt Strauß neuerdings gut" (where "schmeckt gut" = "likes the taste of", "neuerdings" - "recently", "Strauß" - a) the leader of the CSU, b) ostrich) for instance the choice of reading "Kohl = the leader of the CDU" results in either "schmeckt" having to be interpreted metaphorically or "Strauß" as "meat of the bird ostrich" (conversely an analogous form applies if the reading "Strauß = the leader of the CSU" is chosen). For the two sentences "Sie telephoniert" (She is telephoning) and "Er ruft die Kellnerin" (He is calling the waitress) for example a different order of interpretation leads to different results. Fourthly and finally linguistic semantics needs a concept of intension; the model-theoretical concept of structure in its usual version, however, is oriented exclusively towards extensional meanings and the Montague concept of intension based on it lacks empirical plausibility as it operates over possible worlds. All the four defects mentioned can, however, be resolved in a very simple way, only so far no one has thought of doing it like this. I suggest the following modifications. A *situation* S consists of three components: a *universe* X, a *focussing interval* T and an *interpretation function* I; compared to the familiar concept of structure the only thing that is new here - from a formal viewpoint - is in the first instance the component T. As elements of the universe X one can imagine any structures whatsoever (this time in the usual mathematical sense, that is without relation to an object language); for an empirical interpretation it is additionally intended that linguistic structures, mental structures and other object constellations of reality occur in X. The focussing interval T is a totally ordered set; one can imagine it as an interval of time or as a set of points in time. With the help of X and T *focussing sequences* are formed, these being functions from a finite subset of T to X. The interpretation function I is a ternary function operating over X, T and the set of focussing sequences. Following the intended empirical interpretation therefore not only linguistic objects can be interpreted in S by I but also e.g. objects in the perceptual area. The interpretation's dependence on points in time makes it possible to represent language - dependent as well as language - independent changes in categorisation. The planned dependency of the interpretation on focussing sequences makes available a particularly important descriptive instrument: the interpretation of an object takes into account which objects were previously focussed on and - it is implicitly assumed - interpreted. In Kindt(1982 b and 1983) a I have expounded how one can explain the Sorites Paradox and the appearance of so-called hysteresis phenomena in general in a particularly simple way by means of the concept of focussing-dependent interpretation. More generally, however, the focussing dependency is required to deal with interpretations of mutual dependency and the problem of order. Apart from the condition that the interpretation I (x, t, f) of an object x at a point in time t relative to the focussing sequence f is only defined if t is a point in time later or the same as the last point in the focussing sequence, for the present no further limiting conditions are imposed for the interpretation function. In particular nothing is said about what values I assumes and where these values lie. For the intended empirical interpretation, however, it can be made concrete that the values of I can belong to each of the three above-mentioned types of objects or classes of objects of these types or even appropriate combinations thereof. In particular one may imagine for an interpretation of linguistic expressions that it comprises a phonological and syntactic categorisation, the assignment of a mental structure as intension and if necessary the assignment of an extensional meaning. Insofar then as for example the interpretation of expressions like "know", "believe" etc. operates over intensions, i.e. mental structures, the well-known problems of substitution with these (intensional) expressions solve themselves automatically. I cannot, however, enter into a discussion here of these and many other problems which are easily dealt with in the framework now provided (cf. however Kindt 1986). With regard to the topic of "Dynamics", however, I should just like to demonstrate with a few examples how the treatment of meaning flexibility is made possible by the concept of focussing dependency.

3.3 The example sentence already discussed in 3.2 "Kohl ißt und redet Kohl" can be represented as a linguistic structure in the form of a sequence:

w = <0, *Kohl*>, <1, *ißt*>, <2, *und*>, <3, *redet*>, <4, *Kohl*>

If one leaves aside a discussion of the question what principles hold for the employment of focussing strategies one can establish as a possible focussing sequence for w:

$$f = \langle t_o, w_o, \rangle, \langle t_1, w_1 \rangle, \langle t_2, w_4 \rangle, \langle t_3, w_2 \rangle, \langle t_4, w_3 \rangle, \langle t_5, w_4 \rangle$$

Relative to every section of f containing $\langle t_1, w_1 \rangle$, $w_o = \langle o, \text{Kohl} \rangle$ can always have the reading "Kohl = the CDU leader", whereas $w_4 = \langle 4, \text{Kohl}\rangle$ receives varying interpretations: relative to the sections reaching to $\langle t_1, w_1 \rangle$ or $\langle t_2, w_4 \rangle$, w_4 has the reading "Kohl = vegetable" for points in times lying before t_4, but later relative to corresponding sections of f generally the reading "Kohl = nonsense".

With a sentence like "Kohl ißt Herr Schmidt besonders gerne" (= Mr. Schmidt particularly likes to eat cabbage) use can be made of the instrument of focussing dependency to deal with the possibility that up to the processing of "Kohl ißt" "Kohl" receives the reading "Kohl = the CDU leader", but that a reinterpretation of "Kohl" as "Kohl = vegetable" becomes necessary once "Herr Schmidt" has been interpreted.

The examples discussed so far are all to be found in the field of choice of reading for ambiguities. Another important use of semantic flexibility is given with the phenomenon of the relativity of meanings. In a sentence like e.g. "the mouse Mickey is big and the elephant Jumbo is small" the adjectives "big" and "small" are interpreted relative to the size standards of mice and elephants; thus the simultaneous truth of the sentence "Jumbo is much bigger than Mickey" will produce no contradiction. It should now be easy to see that meaning relativity can be reconstructed as a phenomenon of focussing dependency. Ambiguity and relativity also have in common the fact that they lead - in a mathematically topologically definable sense - to incontinuous changes of reading: even 'small' focussing steps result in large changes in meaning. By contrast the vagueness of linguistic expressions offers the possibility of carrying out continuous processes of meaning changes. Using the concept of focussing-dependent interpretation a state of affairs can be modelled whereby in principle the interpretation of every linguistic expression is minimally modified in retrospect by the effect of subsequent interpretation of other expressions. From experiences with conversation-analytic investigations this modelling aspect seems to me to be not just theoretically desirable but also empirically necessary. In Kindt 1982a I have developed a theoretical basis for the variety of meaning changes which are made possible by the vagueness of linguistic expressions; essentially this variety comes about by the meanings of underlying distance measures or uniformities (cf. e.g. Schubert 1964) being modified. From an empirical point of view all those facets of meaning transference belong in this area of meaning change which play a central part in communication and whose importance has in my view been greatly underestimated by semantics so far. The following example may serve to illustrate the type of change based on vagueness. "Wehner is old, Brandt is old, Schmidt is old, even the former leader of the Jusos Schröder - you only have to look at his face - is already old". With the interpretation of "old" changes successively: that is, for a reader knowing the approximate age in years of the politicians named it shifts on the one hand in a direction where younger and younger persons may be called old; on the other hand the meaning aspect "aged appearance, wrinkled face" is introduced into the interpretation implicitly by recalling the appearance of the four politicians and explicitly by the addition of "you only have to look at his face". The games with meaning changes is also impressively demonstrated by the investigations of W. Wildgen (1982) for the example of composition processes.

Apart from the focussing on linguistic objects underlying the above examples interpretation regulation can of course also occur e.g. because of parallel perception of nonverbal actions and events or because of the memory of past states of affairs etc. The differences in the focussing of constellations in the perceptive area and in the activation of knowledge are very often the reason for differences of interpretation between participants in communication.

3.4 The concept of situation put forward here, which can be used equally for the modelling of individual and supra-individual meaning assignations, represents of course only a general theoretical framework and does not remove the necessity of empirically investigating the dependencies of interpretation particularly of focussings assumed to exist. This again involves both the question what interpretational scope actually exists for a given linguistic object and the question what principles of regulation hold. On investigation of the external, interactive level it turns out - this can already be said - that point for point every linguistic expression can take on almost any meaning at all in as far as communication participants agree interactively on a corresponding meaning. Exploration of the internal level will undoubtedly require much effort yet particularly with the investigation of the principles for the construction of figurative and compositional meanings. To explain the phenomenon of meaning regulation the topological concepts I have described in Kindt 1982a, b should in my view be brought in again on the internal level. I can give only a brief outline here of the justification of this view, as follows. If during the interpretation of an utterance along a focussing the interpretation of an as yet unprocessed expression or the reinterpretation of a provious expression is pending, whereby various meanings may be chosen from, then the meaning selected is always the one with the '*shortest distance*' to the total meaning worked out so far and to the current knowledge; this only applies, however, if coherence is assumed and integration of meaning is being striven for. Appropriate distances can be defined via both the probability of states of affairs (cf. Kindt 1974) or via conversational relevance (cf. Smaby 1979) etc. For a successful modelling of meaning regulation it will hence be necessary to gain information about the empirically underlying distance measures. As regards the proportion of distance values going beyond innate perceptual and cognitive mechanisms and dependent on conventions, one can hope to get important information from the analysis of communication and in particular from language acquisition processes. This remark leads back to consideration of the external level, the task of which is more generally to investigate all socially standardized, explicit interaction procedures of meaning construction and meaning regulation I should like to go into these problems in more detail in the following chapter.

4. Meaning Constitution and Interaction

4.1 Every empirically oriented conception of semantics must take into account the fact that meanings are the results of individual interpretation processes and that accordingly empirical access to them only exists via observation of participant behaviour. At the same time present-day knowledge of the flexibility and diversity of meaning constitution in my view no longer permits one to abstract in a naive manner from reference to the situation and participants and speak, say, of *the* meaning of a linguistic expression. If despite this it is undisputed that understanding is possible between communication participants and in fact is often achieved, then one may not, certainly, assume the identity of the interpretation results of different participants to be the normal case, can however ask how similar the meanings in question are or in what way they are made similar to one another. A provisional, somewhat sophistical answer to the first part of this question is: in so far as a communication proceeds successfully the similarity was at least so great that this success was possible. With this answer a functional dependency of similarity on the communicative objective is hinted at, but unfortunately nothing concrete said about the extent of the similarity in practice. If for example a communication participant A says to a participant B: "Would you bring me the jarn from the kitchen cupboard" and if B complies with this request we are still far from knowing how much the individual meanings of A and B correspond for the word "jarn"; taken to the extreme B does not need to know the word "jarn" at all and is then only able to determine the correct object of reference for the phrase "the jarn" on the basis of his assessment of the situation. One should not, however, let oneself be confused by the discussion of such isolated examples. On the one hand, in view of the automatism and speed of successful understanding it is inconceivable that the members of a speech community constitute very differing meanings for every word and reach a common interpretation only after running through additional interactive assimilation procedures. On the other hand it must be remembered that an essential task of socialisation consists after all in meaning acquisition: here in varied and time-consuming interaction processes meanings are established and stabilized and the use of meaning construction procedures internalised and automized. We are thus quite justified in assuming that considerable agreements exist in the meanings constituted and that, at least for elementary linguistic expressions and syntactic patterns as well as for certain meaning constructions, a socially stabilized repertoire of standard meanings exists for each. In this, however, on the one hand the dependency of situations and in particular of focussings must be taken into account systematically. For this point too it is in my view necessary to take a concept of standard situations or focussings as base (cf. Kindt 1981). On the other hand it must be made clear what is to be understood by "standard". Following the reflections above it is neither logically necessary nor empirically plausible to define "standard" in such a way that on each occasion complete agreement of the interpretation of all communication participants is assumed (this point has become particularly clear to me as a result of discussions with H. Rieser on the approach in Kindt 1981). On the contrary use of the word "standard" can only mean that there is a 'common kernel' to the interpretation, i.e. that an agreement exists for the uses of the category concerned in clear, prototypical cases and that varying classifications are possible in 'borderline cases'. Even the agreement postulate for the common kernel is only to be taken as a statistical assertion; not every member of a speech community need for example have an interpretation for every linguistic expression which is in accordance or even merely compatible with the common kernel of the majority of the other members. Such an exactly-defined concept of standard is very closely linked to the phenomenon of vagueness of linguistic expressions (cf. Kindt 1982 a, b).

Variations in the results of interpretation of different communication participants can be due among others to the following reasons:

- Differences in interpretation in the borderline area of standard meanings.

- Differing selection of standard meanings or standard meaning constructions.
- Assignment of situations to differing standards.
- Use of differing non-standard interpretations or differing non-standard interpretation of situations.

Insofar as such differences form a block to a necessary understanding and this is also noticed by the communication participants they will try at least in points to remove or lessen the differences. With this I return to the second part - question above, by what procedures meanings are made similar to one another.

4.2 I consider the investigation of the interactive procedures for meaning assimilation a central task of empirical research, for which it is still necessary to gain acceptance in linguistic semantics however. In the following remarks I should like to outline some results of my own research efforts (cf. also Kallmeyer/Kindt 1979, Kindt/Weingarten 1982).

To begin with one can distinguish among the assimilation procedures occurring 'prophylactically' employed and 'repairing procedures'. With the use of the first type of procedure a communication participant assumes that he or a communication partner will not understand or was as yet unable to understand a certain utterance; he therefore invitates prospectively or retrospectively a procedure intended to improve conditions for understanding. The second type of procedure is only used retrospectively and presupposes that a communication participant has 'noticed' a problem of understanding in himself or one of his partners and that treatment of the problem is 'agreed' jointly. Owing to the explicitness of the repairing procedure one can fairly quickly indicate a pattern as to which procedural components generally occur or can occur in which distribution of roles (cf. Kindt/Weingarten 1982). In the area of monologue communication interactivity of the assimilation procedures is of course only guaranteed in an indirect sense, on the part of the speaker there is only the possibility of prophylactic meaning assimilation and on the part of the listener it is at present still hard to say if the procedures he employs can be traced back to those in dialogue communication. A systematic and comprehensive determination of the prophylactic procedures will require considerable research effort yet. Here too of course there are some particularly striking procedural forms and ones which are in part already known in linguistics; among these is for example the explication of prerequirements for knowledge. Of particular importance also, however, seems to be e.g. a procedure discovered by W. Kallmeyer and myself of 'successive meaning replenishment'. In general one can perhaps assume that all assimilation procedures are oriented to a 'schema of understanding', whose components represent the tasks to be carried out in meaning constitution (cf. Kallmeyer/ Kindt 1979). Aiming at an understanding to be reached in the most economical way possible only those tasks with the accompanying, typical procedures are worked on - one can imagine - from the schema of understanding whose treatment, relative to the degree of assimilation desired, is absolutely necessary.

In addition to the research task of determining the interactive assimilation procedures one must also confront the problem of what decision criteria are taken as the basis for the choice of direction of a meaning assimilation. One can for instance establish that for slight differences of meaning there is considerable willingness to tolerate or in points to accept the interpretation of other communication participants. In many cases the direction of the meaning assimilations is decided by consulting certain 'instances'. In this sense for example the argument within the philosophy of language between intentionalists and conventionalists proves to be a spurious problem: reference to a speaker's intention or a convention of language are merely two different examples of the empirically occurring possibilities of getting a meaning accepted by reference to an instance. Another possibility for the choice or introduction of an instance consists e.g. in appealing to an authority. Tolerance and choice of instance are moreover to a large extent dependent on the particular interaction conditions, which are however in part only established while the interaction is going on.

In all it should become clear from the above considerations that a large part of the dynamics of meaning constitution only becomes explicable when sufficient knowledge is available concerning the interactive procedures for meaning construction and assimilation.

Bibliography

Ballmer, Th., 1973, *A Pilot Study in Textgrammar.* Kolloquiumsvorlage, Bielefeld.

Ballmer, Th., 1978, *Logical Grammar with Special Consideration of Topics in Context Change*, North-Holland, Amsterdam.

Ballmer, Th./Pinkal, M. (eds.), 1983, *Approaching Vagueness,* North-Holland, Amsterdam.

Ballweg, J./Glinz, H. (eds.), 1980, *Grammatik und Logik,* Schwann, Düsseldorf.

Bäuerle, R. et.al. (eds.), 1983. *Meaning, Use and Interpretation of Language,* de Gruyter, Berlin.

Burghardt, W./Hölker, K. (eds.), 1979, *Text Processing-Textverarbeitung,* de Gruyter, Berlin.

Daum, K. (ed.), 1980, *Fingerzeige für die Gesetzes- und Amtssprache,* Verlag für deutsche Sprache, Wiesbaden.

Ebbinghaus, H.-D. et al., 1978, *Einführung in die mathematische Logik,* Wiss. Buchgesellschaft, Darmstadt.

Eikmeyer, H.-J./Rieser, H., 1979, "A Formal Theory of Context-Dependence and Context-Change", in: Th. Ballmer/M. Pinkal, 1983, pp. 131-188.

Eikmeyer, H.-J./Rieser,H., 1981, *Words, Worlds and Contexts,* de Gruyter, Berlin.

Eikmeyer, H.-J./Rieser, H. (eds.), 1982, "Procedural Grammar for a Fragment of Black English Discourse", Ms., Universitätsschwerpunkt Mathematisierung, Universität Bielefeld.

Guenthner, F./Schmidt, S.J.(eds.), 1979, *Formal Semantics and Pragmatics for Natural Languages,* Reidel, Dordrecht.

Kallmeyer, W./Kindt, W., 1979, "Die Untersuchung von Verstehensproblemen", in: *Zwischenstation Universität Bielefeld,* Kommissionsverlag Pfeffer, Bielefeld 1979, pp. 92-97.

Kindt, W., 1974, "Ein Versuch zur Analyse und modelltheoretischen Beschreibung von Rezeption und Interpretation", Kolloquiumsvorlage, Bielefeld, in: W. Burghardt/K. Hölker, 1979, pp. 140-162.

Kindt, W., 1979, "Ziele, Probleme und Leistungen logischer Analysen in der Linguistik", in: J. Ballweg/H. Glinz, 1980, pp. 154-173.

Kindt, W., 1981, "Word Semantics and Conversational Analysis", in: H.-J. Eikmeyer/H. Rieser 1981, pp. 500-509.

Kindt, W., 1982a, "Two Approaches to Vagueness: Theory of Interaction and Topology", in: Th. Ballmer/M. Pinkal, 1983, pp.361-392.

Kindt, W., 1982b, "Neue modelltheoretische Ansätze für die Semantik", in: R. Bäuerle/A. v.Stechow, 1983.

Kindt, W., 1983a, "Dynamische Semantik", in B. Rieger, 1985,pp.95-141

Kindt, W., 1983b, "Zur interaktiven Behandlung von Deutungen in Therapiegesprächen", in: *Journal of Pragmatics* 8 (1984), pp. 731-751.

Kindt, W., 1986, "Die empirische Erforschung von Bedeutungen. Zur Konzeption und Methodologie der Dynamischen Semantik", in: Petöfi et.al. 1986.

Kindt, W./Weingarten, R., 1982, "Verständigungsprobleme", in: Deutsche Sprache 3/1984, pp. 193-218.

Rieger, B. (ed.), 1985, *Dynamik in der Bedeutungskonstitution*, Buske, Hamburg.

Petöfi, J.S. et.al. (eds.), 1986, *Research in Text Meaning. A Survey*, Buske, Hamburg (to appear).

Schubert, H., 1964, *Topologie*, Teubner, Stuttgart.

Smaby, R.M., 1979, "Ambiguous Coreference with Quantifiers", in: F. Guenthner/S.J. Schmidt, 1979, pp. 37-76.

Wildgen, W., 1982, "Makroprozesse bei der Verwendung nominaler at-hoc-Komposita im Deutschen", in: *Deutsche Sprache* 3/1982, pp. 237-257.

LANGUAGE ACQUISITION IN AN INTERACTIVE; SEMANTIC-COGNITIVE, AND NEUROLINGUISTIC PERSPECTIVE*

Ton van der Geest
Ruhr-Universität Bochum

Abstract

Two conflicting theories with respect to the processes underlying the acquisition of language (viz. the language acquisition theory (L.A.) relying on the 'innateness hypothesis' and a language learning theory (LL) relying on general learning and mother-child interaction principles) will be discussed on the basis of recent literature and linguistic, psycholinguistic, and neurolinguistic evidence. It will be argued that there is a great deal of evidence supporting the latter (LL) theory at least insofar as the earlier stages of language acquisition are concerned. Finally after having concentrated upon the fact that language develops in the child almost unnoticed we will discuss some implications for both language acquisition research and language education.

I. Two conflicting theories of language acquisition

Literature focusses during the last two decades mainly upon two conflicting theories with respect to the underlying processes of language development. The first theory - which I will call language acquisition (LA) - is based upon Chomsky's innateness hypothesis. It emphasizes that the child creates his own linguistic rules rather than deducing them directly from the primary linguistic data the child is confronted with. In this theory the argument that in children's speech peculiarities occur that cannot be traced back to the adult system plays a crucial role. See e.g.

---

* A revised and enlarged version of my "Language acquisition as a hidden curriculum"; in: *Communication and Cognition, 1974 (7), 169-190.*

Menyuk (1969), McNeill (1970), Gruber (1967), Wode (1978), and Felix (1978). The other theory - language learning (LL) - is based on the view that the primary data are sufficiently richly structured that the child is able to deduce the linguistic system from these data.

In the first theory (LA) both semantics and the linguistic environment are of secondary interest, in the alternative LL-theory they are - in cooperation with the child's learning ability - essential. In the LA-theory it is apparently sufficient to describe the rules of the grammar accounting for the child's utterances: in the alternative LL, however, a systematic comparison between the input and the output data is mandatory.

Obviously the deviations of child language from the adult model are of primary importance in dealing between the two alternative theories. If it can be demonstrated that these ungrammaticalities are a) merely linguistic, b) systematic and c) unrelated to the adult's model, then they can be taken as arguments pro LA. If, however, it can be demonstrated that the differences between the child's and the adult's system are not differences of code but either of underlying cognitive or physiological and neurological limitations and that they, furthermore, can be explained from the input data, then it seems reasonable to take them as an argument for LL.

I.1. The language acquisition (LA) theory

Studies in the recent past on child language - mainly inspired by the writings of Noam Chomsky focus generally upon the first theory mentioned in the preceding.

In this theory it is stated that the child is equipped with so-called innate ideas with respect to the general characteristics of the grammar of a natural language. These characteristics, which are assumed to be universal, can be subdivided into formal and substantive universals. The formal universals are concerned with the formal aspects of the grammar (e.g. its rules, rewriting, transformational rules, etc.) the substantive universals with such categories and subcategories as nouns, verbs, colour names, etc. The assumption of universals are a result of the limited hypothesis space of the language acquisition device (LAD). Because this limited hypothesis space is given from the onset, early language should be largely universal across languages and cultures (Levelt, 1973, see also Slobin (1970) for discussion). It should be added that LA-theory is only concerned with the acquisition of grammar and not with the acquisition of such communicative aspects as registers and speech styles, neither with pragmatic (illocutive or perlocutive) aspects, nor with interactive (e.g. turn taking and giving) aspects. Therefore, it is concerned with knowledge of language that all speakers share, and not with aspects that differentiate between people. Because of the innate ideas the child is assumed to possess some general expectations about the language to be learned; he is equipped with a general language acquisition theory. This way we can for example expect the child to be able to distinguish between proper or grammatical sentences and improper or ungrammatical sentences. We can base such an expectation on Chomsky's belief that the child is confronted with a so-called informant-presentation of the language to be learned, by which representation all kinds of grammatical and ungrammatical sentences in the child's linguistic environment are enumerated. Or in Chomsky's words (Chomsky, 1965 p. 201): "... it seems that a child must have the ability to 'invent' a generative grammar that defines well-formedness and assigns interpretations to sentences even though the primary linguistic data that he uses as a basis for this act of theory construction may (...) be deficient in various respects." Because the child arrives ultimately at the capacity to produce the grammatical sentences of the language of the environment and not the ungrammatical ones, he has according to LA-theory to rely only upon the grammatical productions of the environment.

As a conclusion it might be said that the child's role in language acquisition is just a matter of mastering the local manifestation of language. From the preceding it will be evident that research in language acquisition within LA-theory will concentrate upon the analysis of output data, the child's linguistic productions, in order to come to grips with the LAD. Actually the only incidental cases in which primary linguistic data and output comparison was made are those to argue that children 'create' rules they never could have found in the data, because they would lead to ungrammatical productions in the adult system.

McNeill (1966) for example argues that such occurrences as 'a gas here' and 'allgone lettuce' cannot be observed by the child from his environment and that, therefore, the child must have created by himself some rules accounting for these peculiar occurrences.

Gruber (1967) argues that children rely in their earliest productions upon the topic-comment distinction rather than upon the subject predicate distinction, because of frequent postposition in the data of the noun with subject function. According to Gruber this topic-comment distinction must be supposed to be innate because the parents did not apply it in their productions.

A last point of LA-theory that is relevant here is related to the assumption that this theory is only concerned with the growth of syntax: it is purely syntactic input which leads to the inference of syntactic structures. Chomsky's LAD is therefore nonsemantic in nature. The only role of semantics in the language acquisition process is according to him that it provides the motivation for language learning while playing no necessary part in its mechanism (Chomsky 1965; chapter 1).

Wode & Felix (1979) argue that certain developmental peculiarities occur in LA irrespective the language nor whether it concerns first or second LA.

I.2. The language learning (LL) theory

The theory - or theories - of language learning admit that the child has certain discovery or learning principles at his disposal but they deny that the child has specific principles for language or its grammar. LL-theories assume not that the child is able to systematize the input but that the input itself is systematically organized to a certain extent. Thus the child is not assumed to be confronted with an informant presentation but with a text presentation, i.e. only correct instances of the model language are enumerated in a specific organization. The LL-theory emphasizes the analysis of the output in its relation to the input. The child's increasing knowledge of the language system to be learned is not only the manifestation of the child's capability but also of the input he is confronted with. Thus only after a systematic comparison of input and output data are we able to state whether a linguistic feature is due to the primary linguistic data or to the LAD in the child.

In this respect the theory can adequately deal with such personal differences as idiolect, registers and speech styles.

Some psycholinguists who have recently supported the LL-theory and emphasized in particular the role of the linguistic input are:
- Brown and Bellugi (1964) who found (in contrast with Bever, Fodor and Weksel (1965) that all adult sentences in the data spoken to language learning children were rather simple and perfectly grammatical, and furthermore, that word order and stress patterns of input and output were identical. To expand the data beyond the special kind used by Brown and Bellugi-children's imitations and adult's expansions of child utterances - the present author replicated this investigation with the same results. See Van der Geest (1972 and 1974) and also further in this paper.

- Schlesinger (1971) supporting the LL-theory claims that language must be learned on the basis of what can be characterized as semantic primacy.

- Snow (1972) found that 1. the speech style of mothers speaking to young children was simpler and more redundant than their normal language use; 2. these changes depend on the reactions of the child to a certain degree; experienced mothers were slightly better in predicting necessary changes in speech style.

Levelt (1975) while discussing Gold's and Horning's studies of grammatical inference in the perspective of LAD argues that:

1. the available evidence is probably more in agreement than in disagreement with the empiricist LL-model, which is according to Levelt essentially also an LAD-model;
2. Chomsky's assumptions with respect to the LAD-model lack support both theoretically and empirically;
3. the consideration of semantic input is a promising and fashionable route to take, although it leads to highly complicated problems in grammatical inference theory;
4. one could consider, as Gold says, the primary linguistic data as an intelligent mode of information presentation, by which is understood that the child is initially confronted with a very primitive subset of sentences which is gradually expanded as his competence grows.

I.3. Empirical evidence

In the following paragraphs we will discuss the results of some experimental investigations in order to demonstrate the relevance of the above mentioned four points, which are mainly based on computer simulations of grammatical inference. Furthermore, we want to arrive at a theory of language development in this section, that can adequately incorporate such phenomena as semantic input and intelligent information presentation.

Brown and Bellugi (1964) - but see also Ervin (1971) - found that the utterances spoken to the child were rather simple and perfectly grammatical. They were almost the kinds of utterances the children used half a year later themselves. They found that the child's spontaneous imitations and the adult utterances were identical with respect to word order and sentential stress. Furthermore, it appeared that non stressed elements, generally function words, tended to be omitted. Child utterances and spontaneous adult expansions were similar the same ways. Furthermore, Brown and Bellugi argue convincingly that children disregard sentences not spoken to them. Because of the rather peculiar status of the utterances used in their investigation (viz. imitations and expansions) I decided to partially replicate this investigation.

The utterances (types, not tokens) of five children to a total of a thousand were provided with paraphrases that were as simple as possible, that were appropriate in the context and the situation, and that reflected as much as possible the language of the adult as it appeared from the data when he talked to the child. The outcomes were strikingly similar to those of Brown and Bellugi. From these findings I hypothesized (see Van der Geest 1972) that the child's limited memory span - both in perception and production - causes the adult utterances to be translated into simpler ones the child can handle. The child's linguistic competence thus grows gradually with his memory span (see also Clark, 1978). Furthermore, it seems that the child's output is semantically rather than syntactically organized in the initial stages (for evidence see below). In Gold's terms we can say that the child is confronted with a text presentation, because the model utterances are perfectly grammatical.

It seems also that the primary linguistic data reflect certain simplicity conditions, which means that the presentation is 'intelligent'. Finally, it can be hypothesized that the child's limited memory span serves as a kind of discovery procedure: less essential, i.e. understressed, material is filtered out. Therefore, we must conclude that a language acquisition theory as worked out in I.1 must be abandoned.

Another type of research on mother - child language that is relevant for the present purpose can be characterized with the term frequency matching. In that research the frequencies of certain structures in mother and child utterances are compared. Bowerman (1972) in a study of the acquisition of Finnish found that the frequencies of different possible orders of subject, object and verb in both child and adult language were quite similar. Brown and Hanlon (1970) discuss the order of emergence of some related sentence types in terms of what they call cumulative derivational complexity. The sentence types investigated were in order of increasing complexity: 1. simple active affirmative sentences; 2. questions; 3. negative sentences; 4. truncated predicates; 5. negative questions; 6. truncated negative sentences; 7. truncated questions; 8. truncated negative questions. It appeared that both complexity and frequency of the input predicted the order of emergence reasonably correctly: Brown and Hanlon argue that frequency and derivational complexity are closely related, and that this relation probably has an independent origin, whatever this may mean.

Frequency and complexity, however, have contrastive value in the formation of theories for language acquisition in this case. Increasing complexity in the child's speech favors the theory that the child internalizes the language system more or less consistently with linguistic theory and favors therefore LA-theory, whereas frequency match favors the theory of LL. which states that environmental data determine to a large extent the progress of the child's development. Brown and Hanlon's results (which seemingly support both theories) are probably essentially in favor of LL-theory; for, because the mother's text presentation reflects derivational complexity, it can be hypothesized that the text presentation is the basis for derivational complexity in child speech. In other words, mother's awareness of linguistic complexity causes a linguistic input for the language learning child that leads to the acquisitional order as found by Brown

& Hanlon. Because Brown and Hanlon's data were not very decisive we replicated their investigation with some modifications (Van der Geest et.al. 1973, 1977); the frequency of occurrence of linguistic variables in subsequent stages of mother and child speech was compared. We found that:

1. semantically the child's speech is more advanced than the mother's speech;
2. syntactically the mother's speech is much more advanced (in comparison with the semantic differences) than the child's speech.

These two points suggest that in mother-child interaction the child determines one way or another how complex the daily conversation is allowed to be in semantic cognitive terms, and that the mother operating within the child's cognitive experience takes the opportunity to provide the child with the correct realization rules to cover the semantics of the conversation.

3. The nonspontaneous utterances of the child are with respect to their syntactic aspects more advanced than the spontaneous ones (spontaneous - first sentence of a conversation).
4. With respect to the semantic aspects the nonspontaneous utterances were less advanced than the spontaneous ones.

3. and 4. are analysed in connection with respectively syntactic and semantic complexity of the mother's speech. The results suggest that mother-child conversation is favorable for language acquisition, for it seems that: a) the mother adapts semantic complexity to that of the child: and b) the child adapts syntactic complexity to that of the mother.

5. There is a strong tendency in the data for spontaneous mother sentences to be less advanced in both the semantic and the realization assessment than nonspontaneous sentences of the mother.

From points 3, 4 and 5 it can be deduced that a mother child interaction starts rather simply and grows more complex. One can hypothesize, therefore, that the conversation between mother and child has a positive influence on the child's language development, especially inasfar as its syntactic realization is concerned.

6. Mother utterances which are responded to by the child are generally less advanced both semantically and syntactically than those not reacted to.
7. If a mother sentence is semantically too simple in comparison with the child's non spontaneous productions then the child will not react.

From 6. and 7. it can be deduced that the adult has to maneuver his sentences with respect to semantic complexity between too complex in comparison with the child's spontaneous productions and too simple in comparison with his nonspontaneous productions.

Our findings suggest that the child's acquisition of syntax derives from two different sources: his own cognitive semantic abilities and his mother's 'intelligent' provision of syntactic information about how to express his progressively more complex ideas. Note, that these two factors were mentioned by Levelt with respect to grammatical inference research. Furthermore, it supports the idea of semantic primacy as formulated by Macnamara (1972), Schlesinger (1971) and worked out for example by Slobin (1973).

Semantic primacy does not only mean that semantic maturation comes to prevail over syntactic maturation, but also that the former is a necessary condition for the development of the latter (new contents occur for the first time in old forms) at least inasfar as the initial stages of language acquisition are concerned.

It appeared also in the data discussed above that semantic development can be described in a natural way because structures with simple semantic descriptions are mastered by the child before the more complex ones.

This could imply that the child does not so much develop his cognitive semantic capacity in the initial stages on the basis of conscious external influences of the adult but rather upon an internal strategy. Some observations of Bruners' (1975), however, may indicate that these semantic abilities may be trained already in mother-child interplay before language actually starts to develop. Also some observations of Snow (1977) seem to support the assumption that semantics is learned, although for more than one reason her analysis is questionable (see Van der Geest, 1980, in prep.).

In sum, the following conclusions seem to be relevant for the present purposes:

1. the child is equipped with a limited memory span which has a filtering function in language acquisition; that is to say: the memory span makes the already simple input still simpler;
2. the semantic cognitive aspects are probably learned before their corresponding syntactic aspects on the basis of interactive principles. This semantic primacy is of primary importance for the acquisition of communicative skills in the first stages of LA;
3. the primary linguistic input must be considered an 'intelligent' (i.e. adapted to the child's language learning capacity at a certain moment) text (i.e. only positive or grammatical) presentation including the following features:
   a. the adult to child speech is rather simple in comparison with adult to adult speech;
   b. adult to child speech is not progressive in comparison with child speech insofar as its semantics is concerned; that is to say it falls generally within the reach of the child's cognitive semantic development;

c. adult to child speech is syntactically progressive in comparison to child speech; that is to say its syntax is step by step more complex than that of child speech (see also Shipley, Smith and Gleitman, 1969, and Snow, 1972).

From these points it is clear that language acquisition is partly determined by the primary linguistic input (see point 3.) and partly by the child's capacity to learn (see point 2.). There is no evidence available that the child's capacity to learn is specifically organized for language acquisition; point 1. has to do with a general psycholinguistic and probably neurophysiological phenomenon; and point 2 has more to do with cognition than with language.

Therefore, it seems that the empiricist model of language learning is probably more in agreement with the available evidence than the rationalistic model of language acquisition.

II. Syntactic vs. semantic strategies

There is another standpoint - alternative to the one sketched in I.3. - possible by which both theories can be considered as complementary to the extent that they are sequentially ordered in time.

To demonstrate this we must first deal with a study of Thomas Wasow (1973) on the innateness hypothesis (I.H.).

In order to prove the correctness of the I.H. Wasow claims that the existence of strong linguistic universals - i.e. universals that are reflections of specific linguistic abilities and may not be reflections of cognitive abilities (see also McNeill, 1966) - is a necessary condition for the innateness hypothesis. To illustrate this, he works this condition out with the grammatical relations, which he assumes - although incorrectly - to be strong linguistic universals. Wasow inclines to the idea that grammatical relations are determined linguistically rather than cognitively, because such cognitive semantic aspects as agent, object, action, etc. are expressed by grammatical relations even in cases where they are clear from selection restrictions, context, etc. It is, however, false to assume that redundancy in the sense meant by Wasow conflicts with the underlying cognitive determination of subject, object, verb, etc. The only acceptable argument would be that grammatical relations and cognitive semantic relations conflict with eachother or are independent from eachother.

After having discussed Schlesinger's (1970) study on Israeli Sign Language (ISL) - in which it appeared that the addressee of an isolated ISL-sentence is not able to interpret what is the subject and what is the object, Wasow concludes that ISL is not a 'natural human language'.

Furthermore, after having quoted Jane van Lawick-Goodall (1971) ("One of the major differences between man and his closest relative (the chimpanzee: T.v.d.G.) is, of course, that the chimpanzee has not developed the power of speech ...") Wasow suggests that the mechanisms underlying verbal behavior are unique to man, and furthermore, that the properties of language which are reflections of those mechanisms would qualify as strong universals. Therefore, Wasow assumes that the language of chimpanzees, but also of old written Chinese and ISL, because of their fairly great dependency on context, are not natural languages. It is in verbal behaviour that language is contextually independent to a large extent.

Before going into more detail the following methodological remark should be made. One of the essential restrictions linguists like Wasow generally make - especially those working within the framework of extended standard theory - is that they only deal with language inasfar as it is structurally determined. The data, therefore, are not utterances, but sentences and professional intuitions about these sentences. One of the essential consequences of such a restricted competence research is that it abstracts from context, instead of demonstrating that language is independent from context. If our concern, however, were with the utterances of a natural human language (as in the case of old written Chinese, ISL, the language of the chimpanzee, and the language of the language learning child) then we can demonstrate that the context is a necessary condition for correct interpretations of utterances in many cases (see L. Bloom (1970) for discussion). If we accept, inspite of this methodological remark, Wasow's conception of a 'natural human language' then the following characteristics of child language are relevant to the present purpose:

1. the child's language is telegraphic in nature with respect to syntactic realization. In a telegram 'komm morgen' (lit.: come tomorrow) is ambiguous and can be interpreted as 'you must come tomorrow' or as 'I'll come tomorrow'. The same goes also for child language, in which 'Jan slaan' (lit. 'John hit' (infinitive) can be interpreted as 'John is hitting/will hit/ etc. someone' or as 'someone hit/will hit/etc. John'.
2. as was argued in the foregoing sections the child's language relies on cognitive semantic strategies rather than on syntactic ones.

These two points would lead to the assumption that the child's language is not a natural human language at all, and that therefore, the innateness hypothesis is of no relevance for the research of the child's language learning. However, it should be admitted that after a certain period of language acquisition syntactic strategies come to prevail over cognitive semantic ones. This latter point can be illustrated from two studies reported by Bohannon and Friedlander (1973) on selective listening in Kindergarten and elementary schoolchildren. It was found:

1. (Rileigh and Friedlander, 1971) that children until in the second grade do not express a listening preference for either one of two ongoing versions of the same story: one with normal word order (within constituents) and one with randomized word order.
2. If a flat intonation is used in the narrative with normal syn-

tax and a lively intonation in the narrative with randomized syntax, then intonation seemed to serve as an aid to select the narrative with normal syntax (intonation caused a listening preference of 60 percent for syntactically correct narrative).

3. Children from grade two to grade 5 preferred at a high level of significance the monotone meaningful syntactically correct narrative.

From these studies it seems clear that children rely on syntactic strategies essentially only after age 6 or 7.

The following studies are of special importance in the light of the present discussion, viz. those of Strohner and Nelson (1974), Maratsos (1974), and of Bohannon and Friedlander (1973).

Strohner and Nelson investigate the child's sentence comprehension and more especially the influence of the knowledge of the word (or event probability), of the non-verbal context, of the semantic structure and its hypothesized actor-action-object order, and of the syntactic form upon the child's comprehension of reversible and irreversible actives and passives. What they found was that three years old children consistently applied other than syntactic strategies which led to many errors of comprehension. The strategies used were based upon knowledge of the world (event probability) or upon semantically based judgements of word order. By age five the children appeared almost always to rely upon syntactic strategies; only in cases of improbable passive sentences this strategy didn't work satisfactory according to the authors. In my opinion, one can only be certain that a syntactic strategy is working if we find out that it works where it has to, viz. in improbable passives. For all the other instances 'event or probability' and the semantically based 'word order strategy' can account for the correct scores. Therefore our conclusion has to be that syntactic strategies work only very partially at age five.

Maratsos (1974) reexamines how children understand infinitival complement clauses without subject. The results strongly favour the conclusion that preschool children rather make use of a 'semantic role' principle than of a surface structure 'minimal distance' principle. Furthermore it appeared because of a somewhat better performance of the older age group that a syntactic strategy is developing.

This suggests that the rationalistic L.A.D.hypothesis that language development is syntactic as worked out by Chomsky may work after age 7. If this is the case, then the developmental sequence in language learning runs as follows:

A. the child starts his linguistic career with the aid of cognitive semantic strategies and an intelligent or programmed input (empiricist model).

B. It is theoretically possible either that the child masters a sufficient amount of syntactic knowledge from the input such that he is able to exploit syntactic hypotheses at a certain age, or that he masters a certain amount of semantic cognitive knowledge that is sufficiently rich to allow him to switch to a syntactic and possibly innate strategy. Empirical research has to decide between these two possibilities.

C. After a certain age a child can be supposed to exploit syntactic strategies, if and only if it can be demonstrated that these strategies work independently of cognition. The difference with the preceding B. is that in B. there is no mismatch between syntax and semantics. Such a mismatch as formulated in the definition of a strong linguistic universal is essential for stage C. Crucial for the correctness of the standpoint as expressed in C. is the question whether the theory of generative semantics or the theory of interpretive semantics turns out to be adequate. If the latter would turn out to be adequate then Chomsky's LAD-hypothesis can be supposed to be correct after age 6 or 7.

III. Language acquisition in a neurolinguistic perspective

Although the research area of neurolinguistics is really on the march, most disciplines dealing with language neglect neurolinguistic aspects largely in their theoretical considerations. Most linguistic, sociolinguistic, and remarkably also psycholinguistic theories present explanations for their observations without taking notice of neurolinguistical and biological possibilities or impossibilities. The same goes also for the study of language acquisition. This is the more astonishing when we realize that the maturation of the brain, its modifications as for example its lateralisation, may influence essentially the language developmental processes (as for example worked out by Lenneberg, 1967).

On the other hand, neurolinguistics - when dealing with developmental aspects of the brain in relation with language - abstracts largely from developmental aspects with respect to language ability but concentrate on language ability as a static phenomenon. The measurements were generally carried out with aphasic patients in dichotic experiments with children up to the age of 14;0 without giving too much consideration to language acquisition processes (see e.g. Segalowitz & Gruber (eds.), 1977). Furthermore it should be noticed that language capacity is tested perceptively, although one knows that such tests do not differentiate between patients, since amongst others the neuronal apparatures for language perception - in contrast with that of speech production - is rather diffuse and bilaterally organized.

Neurolinguistic results with respect to maturation of the brain (lateralization) can be summarized as: cerebral dominance of the left hemisphere exists probably already since birth. Its effects on language behavior expands with age until the process of lateralization becomes irreversible in puberty (Lenneberg, 1967).

Some explanations are:

- the cerebral dominancy of the left hemisphere expands. By this process the specificity for language in the left

hemisphere increases,
- the plasticity of the right hemisphere for language decreases,
- the left lateralization exists from the beginning, what has developed is a cognitive skill and the possibility for its neural lateralization (Moskovitch, 1977; Wittelson, 1977).

As we may conclude from the preceding discussion the results and their interpretations in developmental neurolinguistic research are rather disappointing. Lateralization certainly exists for language, but when it starts, and (or) when it becomes irreversible and how such a process runs is still largely uncertain.

In my opinion we need a theoretical framework that allows us to transform observed processes in language development into hypotheses with respect to neurolinguistic research. The telegraphic speech style as discussed in section I,3, and II, and its hypothetical correlate in memory and therefore in the brain may serve as an example for the sketched constellation.
Another example may be the developmental phenomena as they occur in the child's language from age 5;0 till 8;0. We mentioned in section III the investigations of Bohannon & Friedlander, Strohner & Nelson, and of Maratsos already. They all suggest that children gradually develop a syntactic strategy, start to control them only after age 5 and start to exploit them after age 6 or 7. Bohannon & Friedlander found that children after age 6 preferred syntactically (left hemisphere; LH) correct narrative instead of a lively intonation (right hemisphere; RH). The other investigations presented also evidence that syntax (LH) comes to prevail over other interpretive phenomena like e.g. contextual, and situational judgement, common sense, event probability, semantic aspects which can all be characterized as localized in the RH and as analogous, not explicitly rule governed.

To complicate this issue, however, there are a number of investigations to be mentioned from which it appears that this rigid syntactic strategy to overestimate formal matters and to neglect accompanying more subjective but essential and substantive matters only works for a few years and disappears at least in its strongest form when the children are 9 or 10, and exploit more pragmatic technics again for the perception of language. Balfour (1975) proves this to be the case for the full passive, Schöler (1979) for affective intonation, Müller-Abé for the comprehension of idioms, and Pechmann (1979) for referential gesturing.

We have to conclude, therefore, that children in the age span from 5 to 8 suddenly give prominence to left dominance linguistic features, and this not only in speech perception but also in production, as they acquire the syntactic aspects of more complicated sentences in this stage like adverbial and complex sentences, passive sentences, complex tense, and irregular morphology. The child acquires also the capacity at this age to adapt syntactically his speech to the linguistic level of his partner's in a flexible way (Van der Geest & Heckhausen, 1979). If we take also into consideration that children at this age change their cognitive strategies from preoperational to concrete operational, and that remarkable decreases in cognitive performance can be observed in this age span (Drubbel, 1976; Koster, 1975; Bolten, 1975), we may hypothesize that all these data must have their correlate in the brain.

In my opinion we have to take into account the possibility that the process of the so called irreversibility of linguistic lateralization must be assumed to be a multi stage rocket that more or less corresponds with the child's cognitive-linguistic developmental stages.

Lenneberg's hypothesis, in which it is stated that the lateralization is a progressive process that is completed at the age of puberty and that before this time the right hemisphere may play its role in language development, has to be reformulated such that the observations as dealt with can be accounted for. If we consider Lenneberg's evidence (tables and diagrams) we have to admit that brain development largely takes place before the age of 7. Finally some other results make clear that Lenneberg's assumption that language learning can only take place between the age of two to about thirteen, must be reconsidered.

Firstly Snow & Hoefnagel (1978) found that older learners (after puberty) have an advantage (rate of acquisition) over younger learners in acquiring the rule governed aspects of a second language - morphology and syntax. Tervoort (1972) demonstrated that deaf children after puberty acquire the essential aspects of the full passive. At age 13;0 he found 30%; at age 14 50%, and at age 15 70% correct answers.

IV. Language acquisition and cognition

The actual problem we are concerned with when we confront Chomsky's language acquisition theory with cognitive psychology is that in a way the child mastering his language can be compared according to his theory with a linguist writing a grammar of that language (see also McNeill 1966). Although Chomsky formulates it carefully by saying that his discussion of the topic "does not refer to conscious formulation and expression of these (rules, hypotheses, etc.) but rather to the process of arriving at an internal representation of a generative system which can be appropriately described in these terms" (Chomsky, 1965, p. 46), the present comparison has been taken too literally many times in actual research. Chomsky's above quoted remark, however, is vague in the sense that it either emphasizes that the linguist's and the child's activity differs in terms of resp. conscious and unconscious or it states that the processes of arriving at an internal representation of a generative system can be described both for the child and the adult in the same terms.

It is clear that a linguistic theory is only explanatorily valid if the processes as meant by Chomsky are appropriately described, that is to say are also cognitively validly described. The very question now is whether one can maintain that the processes - whether conscious or unconscious - of the child's language learning are identical to those of an adult and professional linguist describing a language or in other words that the child's discovery procedures are identical to those underlying the linguist's professional and systematic linguistic intuitions.

If not, then Chomsky's linguistic theory is only psychologically valid inasfar as it describes correctly the ultimate products of different kinds of psychological operations.

There is a lot of evidence that children and adults differ intellectually with respect to thought and language, illustrative to this is Piaget's subdivision into:

a) sensory-motor stage (age 0-2)
b) preoperational stage (age 2-7)
c) concrete operational stage (age 7-11)
d) formal operational stage (age 11-..)

Language acquisition takes place largely in the first two stages: linguistic sophistication is restricted to stage d) and is only latently present in most people because of lack of professional training.

The preoperational child's thought is irreversible and attentive to limited amounts of information, in particular the static states of reality. The child is only able to produce correct mental images of static situations. He concentrates on states rather than on transformations and kinetic strategy (changes in shape and form and position).

Although the evidence is mainly derived from experiments utilizing the visual channel, it seems plausible that these limitations on the child's intellectual functioning are also valid for the auditory channel as it is not the channel but only the deeper cognitive perception mechanisms that concern us here. If it is true, then we can expect in each stage different strategies to operate in the child' treatment of morphology (changes in shape of form) and of word order (position). If Piaget's theory is correct it seems unlikely that children derive one shape or form from the other in the present stage (age 5-7):

- They do have perhaps a shape for tense, plural etc.; they have probably furthermore 3 position rules for verb and affix in the surface structure and for noun and affix, but they fail to apply these rules correctly (Berko 1958) in new instances (e.g. in the case of non-sense words);
- they seem also to have a rule for active and passive sentences, but fail to recognize that different morphology and word order make different interpretations necessary (see e.g. Bever and Langendoen (1970), Turner and Rommetveit (1967), and Balfour (1975).
- they have a position rule (subject-object-verb) but fail to notice that this rule is not valid for certain subordinated clauses (C.Chomsky 1969), F. Kessel (1970)), Maratsos (1974), Strohner & Nelson (1974).

The present problem of how to account for the psycholinguistic processes of the older preoperational child can be illustrated with Berko's (1958) results (see table 1).

	Percentages and means preschool	Correct answers first grade	average increase (percents and means)
plural	14 - 79 (63)	33 - 99 (63)	16,6 (20)
progressive	72	79	25
past tense	0 - 73 (48)	25 - 85 (66,5)	12,5 (18,5)
third p. sing.	47 - 57 (52)	49 - 56 (52,5)	0,5 (0,5)
possessive	74 - 53 (68)	46 - 99 (88,5)	17,5 (20,5)

Table 1: Berko's results on learning English morphology (1958).

From this table it might be deduced that the third person singular which concerns both the NP and VP - is extraordinarily and consistently difficult for both the younger and the older age group. The correct scores of all other linguistic items - which are restricted to either the NP or the VP - increased with age. Thus it can be hypothesized within current Extended Standard Theory, that the child of the present age range becomes able to deal with restricted domain transformations (those concerning a major consistuent of a sentence only) but that he still fails to control the transformations operating on the sentence domain.

Moreover, there are alternative explanations here, that can all be derived from learning theory.

The first possibility deals with associative learning, by which is meant here that a child does have for example a plural article (*the* and Ø) in his lexicon and a plural noun, rather than that he has mastered a rule noun - plural suffix.

In such a view *a* asks for *tooth* and *the* (plur) asks for *teeth*. The child can correctly realize *a tooth* and *the teeth* but he fails to notice that they are somehow or other related to each other. We must assume in such a case, that correct realizations (such as 'a tooth', 'the tooth', 'the teeth', and 'Ø teeth') are controlled by a prior operating cognitive semantic representation of the message (see Schlesinger (1971) and Van der Geest (1974) for discussion). Thus, what one wishes to say controls what one realizes in speech.

A second possibility works out the former one by emphasizing the planning nature of the semantic cognitive component. This can be illustrated by means of / EVENT / sentences or generic sentences (see Van der Geest 1974, chapter 2).
See for example:

(1) Birds sing songs
(2) John hits cats
(3) Peter normally hits his neighbour's cat

In (1) all elements can be qualified as generic. *Birds* means 'for all members of the category of birds': *sing* means something like 'are able to sing' or 'tend to sing' and *songs* 'something that can be qualified as a song'. In (2) only *hits* and *cats* are generic. Because *John* is an identified element (which means that it belongs to the (presupposedly) given part of the analysis) it can be hypothesized that all elements of the comment part of a generic sentence must be qualified as generic. This holds also for (3) if one realizes that it must be paraphrased as 'Peter normally hits a cat which is his neighbour's cat!' The relative clause must be considered here a condition working upon 'a cat'. Therefore, it seems reasonable to state that the linguistic feature *generic* works upon the sentence and that the clements can be specified from high to low that are under the domain of generic. The speaker is at the S-level already aware that he wishes to say something generic or not with respect to a certain state of affairs. This means that by the qualification of S the pre-terminal string is determined with respect to the features plus or minus generic. Something similar can be proposed for such intermediate levels as NP and V or VP, as was argued above for plural.

We must be aware now that the third person singular must be accounted for not at the NP- or VP-level but at the S-level, because it has its consequences for both the NP subject and the V. It seems plausible to suppose that 'planning' at the S-level is more complicated than at the NP- or VP-level because of its greater domain.

Our choice in favor of the present alternative for the child's language at the present state has two important implications.

The first implication is that it can be hypothesized that children in the present stage relying heavily on a very extensive but unsystematic lexicon, have to learn all kinds of morphological rules in order to be able to simplify their lexicon.

The second implication is that we find independent support for the assumption that children in the sensory motor and in the preoperational period rely on a semantically cognitively based language acquisition theory, and that only after age 7 may a syntactically based language acquisition be expected to occur.

The assumption that young children rely on a semantic theory for language acquisition is mandatory if we realize that the semantic information has a steering function (i.e. a planning function) for correct sentence realization, which is adequately accounted for in generative semantics.

Finally, it must be added that semantic theory can also account for the morphological transformations. In this light the discussions between Chomsky and Lakoff (see Wasow for references) are illustrative.

V. Language acquisition: a hidden curriculum

There are three arguments frequently mentioned in the discussions in favor of the innateness hypothesis:

i. Language acquisition takes place in a surprisingly short time.
ii. Language is a very complex system; this makes (i) the more surprising.
iii. Language acquisition occurs almost unnoticed.

Point i. is doubtful. First of all we cannot compare the child's language acquisition period with the adult's second language learning period; for in the latter case the primary language is a confounding variable which on the one hand may accelerate the learning of a second language, but on the other hand produces a certain interference especially at the phonological level. Secondly it is very difficult to determine the exact boundaries of the language acquisition period. Does a child start his linguistic career when he uses his first two word sentence (18months), when he uses his first intelligible word (12 months), when he uses his first intonational pattern of the mothertongue (between 9 and 12 months)? Or when he produces deliberately cries, gestures, and so on to reach a certain effect in the hearer (4 months)? The present question is more interesting with respect to the other bound of the language acquisition period. Is language acquisition finished when the child knows and applies the basic rule system of grammar (around age 5), or when he is able to write complex sentences properly (age 12-14)?

With respect to point (ii), it must be mentioned that the language system is as complex as people think it is, that is to say linguistic theory determines itself to a large extent the complexity of the system.

It can be questioned whether it is that complex system that a child learns and applies when speaking or listening, in other words, whether language acquisition, language production and perception, run according the lines suggested by linguistic theory.

The only reliable argument in favour of LA-theory is, therefore, point iii, that language acquisition occurs almost unnoticed. This qualification 'almost unnoticed' is apparently used in comparison of other kinds of developing behavior such as learning to eat independently and to walk, which are more overly guided gy the adult than language acquisition. LA-theory puts it even stronger by assuming that parental guidance is of almost no relevance for language acquisition. Therefore it is postulated in LA-theory that the processes are organized almost exclusively internally. It is only a small step from this standpoint to the assumption that the child has an innate theory for language acquisition to his disposal.

In the preceding sections I discussed LA-theory and LL-theory especially in the light of the relevance or irrelevance of the primary linguistic data (pld.). What we found was that the pld. were structured and that they therefore, at least facilitate the processes of language acquisition. The central problems now are

a) to determine more exactly the role of the pld. For example, is it essential that the primary linguistic data are structured? Is their only role facilitation of the processes of language acquisition that would occur anyway? If it is essential that the primary linguistic data are structured, as computer simulations makes us believe, to what extent then must they be structured and in what respect must they be structured;

b) to determine more exactly the role of the child's general learning capacity. Do these capacities develop in a way that is important for language acquisition? Is the child capable of formulating hypotheses about language in general and more specifically about syntax? Are these hypotheses similar to the child's general learning principles? Does this kind of hypothesis develop along the same lines as his intellectual development? What do these hypotheses refer to? etc.

In this light it is interesting to remember that LA-theory claims a limited hypothesis space and LL-theory an extensive hypothesis space. This means that in the LL-theory the child is postulated to be more active and creative than in LA-theory;

c) to determine more exactly the child's innate and language specific capacity to learn language. Actually this latter point is nothing more than a methodological gap: all peculiarities that cannot be explained for the moment from a) and b) belongs to it for the time being.

That language acquisition occurs almost unnoticed is only partly true. Every adult is aware that he cannot talk to a child - whether he is one, two or seven years old - in the same manner as to an adult. Furthermore, it is often found that adults don't neglect systematically the child's language difficulties, but do correct and expand their immature productions.

Finally, the analysis of mother to child language makes clear that this kind of communication is bound by very specific rules.

A valid explanation of why people think that children learn their language almost unnoticed is perhaps that language is not learned so much by overt practice in a learning situation, but rather in a communicative situation. What we are doing with a child is communicating, transferring information establishing relationships. For the child, however, there is the additional task of detecting linguistic regularities in the speech directed to him. The difference between language acquisition and the development of behavior that is learned by overt practics and training by the adult is, that in the former one must communicate with someone else on whom one is dependent to a certain extent, which is not the case for walking, eating, etc.

Snow et.al. (1974, ms) work out various kinds of indirect evidence that a simplified speech register - when one speeks to the child - is crucial to language acquisition:

1. children have a simple correct redundant sample of utterances available to them for learning language;
2. children themselves are instrumental in eliciting the simplified register.
3. children educated in a bilingual situation, in which adults address them in only one language, will develop little active control of the language which they only overhear.

All kinds of evidence and arguments in favour of the LL-theory do not only ask for a cognitively and semantically based language learning theory of an interactive nature but also for a 'natural' pedagogical framework.

Such a framework I would like to characterize as 'hidden curriculum'. Let me first describe what I understand by this term and thereafter work it out for research purposes and practical applications.

A hidden curriculum must have a number of characteristics, some of which concern the 'tutor', some the 'pupil' and others both the pupil and the tutor. Some essential conditions are:

i. the teaching of the subject-matter of tuition is a side-effect only of the tutor's task (tutor's role).
ii. the teaching of the subject matter belongs to the ultimate objectives of education (tutor's role).
iii. How much actually is learned is evaluated from time to time by the educator (tutor's role)
iv. the educator will adapt the subject matter to his findings as indicated in point iii. (tutor's role).
v. the child himself is motivated to learn the subject-matter (pupil's role).
vi. the child is therefore actively engaged in the curriculum (pupil's role)
vii. the child himself must detect and process the regularities of the subject matter. This point is accounted for by general learning theory (pupil's role).
viii. interaction between pupil and tutor is essential for the effectiveness of learning e.g. the tutor's reactions function as verifications and falsifications for the pupil (both tutor's and pupil's role).
ix. the subject-matter ought to be chronologically systematically organized; e.g. the child is instrumentally involved in the tutor's adaptation of the subject-matter (both pupil's and tutor's role).

These characteristics, which are also relevant for such hidden pedagogies as discussed by Bernstein (1973) can all be applied to primary language acquisition, second language learning is a natural setting, growing linguistic capacity of the school child, the learning of certain registers such as "the language as used in the schools", etc., and are, therefore, extremely relevant topics for pedagogical investigations.

The relevance of the nine characteristics for language acquisition becomes clear form the following remarks:

(i.a) the child's language learning is a side effect of pupil - tutor communication;

(ii,a) the tutor wants the pupil to speak his mothertongue correctly;

(iii,a) the tutor is aware what kinds of linguistic features are too difficult for the child to process and therefore to learn, as is clear from his avoidance of certain complex linguistic structures;

(iv,a) the tutor simplifies his speech register further when he is aware that he is misunderstood or not understood at all; (Snow, 1972);

(v,a) the child practises linguistic phenomena in non communicative situations (Weir 1962);

(vi,a) the child's reaction to mother sentences are linguistically determined;

(vii,a) this point has been dealt with extensively in section III;

(viii,a) the interaction variables have been discussed in some detail in section I,3. Furthermore, mismatch between tutor structures and those of the child himself will lead to alterations of rules (see also McCawley 1968). Finally, expansions of child utterances by the adults seem to be helpful for the child's mastering of his language (Brown, Cazden, and Bellugi, 1969);

(ix,a) it is evident that speech registers are more simplified for two year olds than for three year olds, etc.

From the nine characteristics of the pedagogical framework and their illustrations in the field of language acquisition research, it is clear that all essential questions about language acquisition dealt with in the preceding sections, fit into the present pedagogical frame. Although this fitting is not essential for language acquisition research it offers the advantage that language acquisition research can be tuned in to pedagogical questions directly, and furthermore, that their findings are directly available for pedagogical workers.

Let us illustrate this briefly with a number of topics that language acquisition research and language education share.

I. Compensation and enrichment programs such as those worked out by Bereiter and Engelmann (1968) didn't have the educational results initially expected. Beside the fact that in this kind of programs specific linguistic features are walked through that the child apparently had available yet (See Van der Geest et. al. 1973, a) these programs neglected the hidden character of language acquisition, and trained linguistic features by overt guidance in a specific 'dry-run' setup.

Actually, work with this kind of program all over the world can be considered as one enormous experiment ultimately leading to the falsification of the hypothesis that verbal interaction - i.e. spontaneous language use directed to the exchange of information by both partners of the interaction - is not a necessary condition for the child's language acquisition. This condition is fulfilled in the orthopedagogical method for language learning known as the natural, free, or Belgian method, which is used in certain schools for the deaf (see also Tervoort, 1953). Illustrative of this method is A. van Uden's (1970) maternal reflective method.

II. There is evidence that verbal interaction between mother and child differs systematically with respect to linguistic complexity in different situations (see Snow et al. 1974). This is probably the case for both mother and child language. It is, therefore, interesting to determine what kinds of daily situations are most effective for the child's language acquisition. For this purpose we can try to isolate linguistic features that are specific for different situations, and try to determine which structures are first used by the child. If it appears to be the case that the features used by the mother systematically in situation 'X' tend to appear in the language of the child before those used by the mother in situation 'Y' then it seems to be justified to assume that situation X is more directly relevant for the language acquisition process than situation Y.

III. It is interesting to know how long and to what degree verbal interaction is a necessary prerequisite for language acquisition.

One can also put the question as follows: "Is the verbal interaction between teacher and child essential for the language progress of the child and especially for the child's learning of the language of the school?". This question is the more important because it is assumed that degree of success in mastering the typical school register determines general school success to a large extent. Therefore, it is interesting to know what kinds of intervening pedagogical variables, e.g. the subject-matter of tuition (language arithmetic, biology, geography, etc.) and what kinds of pedagogical strategies (teacher's monologue, verbal interaction initiated by the teacher or by the child, etc.) are especially important to the child's language progress.

If it appears that the child's language progress is indeed determined to a large extent by the teacher-child interaction, then there are two curricula for language education in the school, viz.

a. the scheduled language lessons, the overt curriculum;
b. the hidden curriculum.

Because it is common practice to evaluate only the overt curriculum, and because this evaluation actually is influenced by the hidden curriculum, it is necessary to come to grips with the hidden curriculum. This is the more iteresting because literature suggests that the hidden curriculum presented by the teacher does not favor all children to the same extent, since teachers generally utilize the typical middle class register which lower class children control poorly.

It is evident that such researchers as Smith (1964) Bellack (1966) Flanders (1970) and Van Trotzenburg (1972), who also aim to describe verbal interaction in the classroom, neglect this language learning aspect of interaction because their purpose is to inventory the kinds and frequencies of pedagogical moves, in order to describe the pedagogical process.

VI. Language acquisition in a pedagogical situation

From the foregoing discussion on communicative development it may have become clear that the rules of communication and the process of communicative development are inseparately related. I regard this revamped 'correspondence'-hypothesis as a hidden curriculum for language acquisition, since the actual task in a parent-child dyad is to communicate and successful communication leads to growth of communicative competence. In another study (Van der Geest, 1978) it is not only stated that communication is a necessary prerequisite for language learning, but that, furthermore, communication is a learning process itself. Inasfar as the participants are concerned we may speak of mental changes with respect to context and these changes, essential for the communicational process itself, may be considered as learning effects. These effects concern amongst others the topic spoken about, the participants, the situation, and, last not least, the code.

The basis for such an assumption are the results of a longitudinal investigation of the communicational processes of a group of children and their teacher from kindergarten to the end of the first grade. The following results are relevant for the present exposure. Initially the four groups of children (lower vs. middle class; boys vs. girls) did not differ qualitatively (more utterances, more words, etc.) in classroom interaction. These quantitative differences occured initially only in the children's verbal productions and were only hesitatingly taken over by the teacher, with respect to whom she spoke. Probably this adaptation of hers was against her assumption of equal participation of the children.

Later the children's participation in classroom interaction also differed qualitatively. In this case, however, it was the teacher who initiated this difference: it was in her speech directed to the different (groups of) children that such differences first came to the surface. The same kind of, or complementary, qualitative differences accordingly appeared later in the speech of these groups of children. We may explain this remarkable communicational phenomenon as follows: the quantitative differences functioned for the teacher (erroneously?) as a signal that there were differences among the children in the classroom. Talking less appearently means for the teacher:talking differently.

Initially we may expect the following to occur:

> Teacher: Nora you may give the answer
> Nora: (hesitates)
> Paul: In the autumn
> Teacher: Paul wait for your turn! Nora?

And now you may guess who gets the next turn. The same type of children who cannot wait for their turns initially are exactly the children the teacher later preferably talks to. And because of the interpretation 'talking much is talking differently' she talks with qualitatively different contents, intentions, and structures to the several groups of children. Ultimately the groups of children develop different registers, namely those that fit into the registers that are directed to them. The children become different native speakers, at least in the different school settings, as the result of an interactive play with the teacher that can be described as a hidden curriculum. Middle class children and boys preferably used an informative and more abstract speech style, and participated most often in the classical conversation of more cognitively oriented lessons. Lower class children on the contrary used a more concrete and more instructional speech style, especially during classical conversation and instruction. The differences were most pronounced in the calculus lessons. The interaction of the teacher with the girls and lower class children was concerned with very concrete action describing instructions of how to solve a mathematical problem. Boys and middle class children, however, got rather abstract, problem oriented and explanatory contents. One can imagine that such a conversational praxis not only leads to differences in communicative competence but also, little by little, to different curricula for mathematics, namely a concrete problem solving oriented one and one that is concerned with more abstract understanding of mathematical problems. It is this latter point that illustrates the close relationship between communication and learning. In figure 1 my concept of communication, learning and communicative development is represented. The double arrows in this diagram indicate that the block to which they point interact with each other. If such an arrow consists of a dotted line then the interaction takes place between the situation and an interaction between two other blocks of the diagram. A single arrow, furthermore, means that the in-

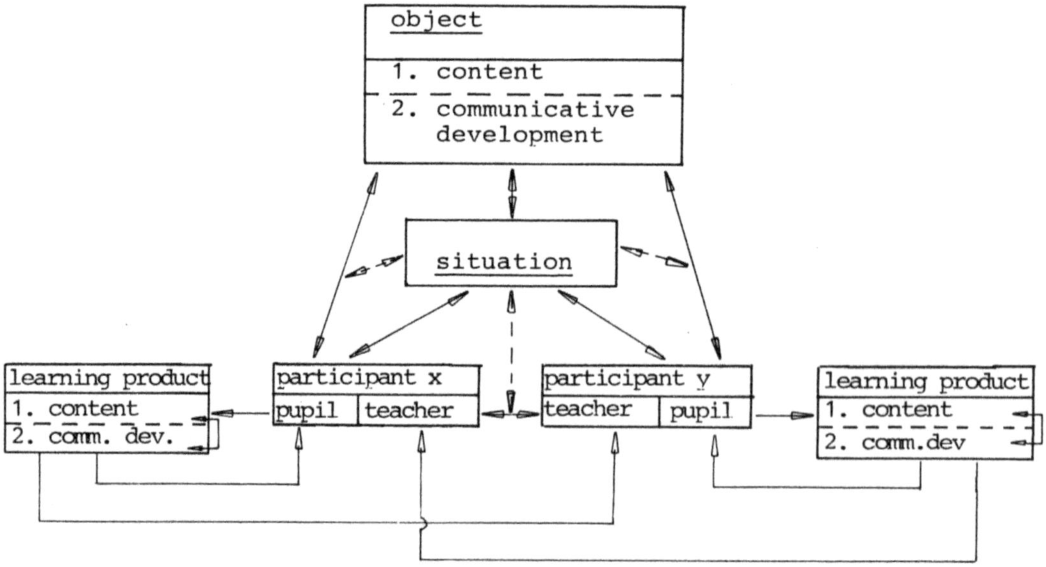

Fig. 1: learning as a communication process (for explanation see the text.

fluence takes place only in the indicated direction. In the two blocks *learning product* and *object* the double function of communication is represented, namely the mental change with respect to both the communicational content and to communicative development. A last point to be mentioned considers the double arrows in the blocks '*learning product*' and '*object*'. After the preceding discussion it will be clear that the communicational object and product not only refers to its content but also to the capability to communicate about these contents. It seems, therefore, reasonable to accept that the communicative competence of the participants restricts or at least influences the contents and vice versa, and that people who developed a different communicative competence with respect to the content will discuss the conversational topics differently.

In sum it can be stated that from recent studies in LA a conception of language learning can be derived that seems highly attractive for those interested in language education and pedagogy and communication in general.

Bibliography

Balfour, G., 1975, *The interaction of syntactic, semantic and contextual information in children.* Paper presented at the 4th International Congress of Applied Linguistics. Stuttgart.

Bellack, A.A., et al., 1966, *The language of the classroom.* Teachers College Press: New York.

Bereiter, C./Engelmann, S., 1966, *Teaching dis-advantaged children in the preschool.* Englewood Cliffs.

Berko, J., 1958, "The Child's learning of English morphology". In: *Word*, 14, pp. 150-177.

Bernstein, B., 1973, *Class and pedagogies; visible and invisible* (MS).

Bever, Th./Fodor, J./Weksel, W., 1965, "On the acquisition of syntax: a critique of 'contextual generalization'". In: *Psychological Review*, 72, pp. 467-482.

Bever, Th./Langendoen, D.T., 1971, "A dynamic model of the evolution of language". In: *Linguistic Inquiry*, 2, pp. 433 -463.

Bloom, L., 1970, *Language development: form and function in emerging grammars.* Cambridge, Mass.

Bohannon, J.N./Friedlander, B.Z., 1973, "The effect of intonation on syntax recognition in elementary school children". In: *Child Development*, 44, pp. 675-677.

Bolten, M., 1975, *Antonieme adjektieven by 5-7jarige kinderen.* University of Amsterdam, Inst.of gen.linguistics, M.A. Thesis.

Bowerman, M.F., 1972, *Learning to talk: a cross-linguistic study of early syntactic development.* Cambridge.

Brown, R./Bellugi, U., 1964, "Three processes in the child's acquisition of syntax". In: *Harvard Educational Review*, 34, pp. 133-151.

Brown, R./Cazden, C./Bellugi, U., 1969, "The child's grammar from I to III". In: J.P.Hill (ed.), *Minnesota symposia on child psychology*, 2, pp. 28-73.

Brown, R./Hanlon, C., 1970, "Derivational complexity and order of acquisition in child speech". In: J.R.Hayes (ed.), *Cognition and development of language.* Wiley: New York, pp. 11-53.

Bruner, J., 1975, "The Ontogenesis of Speech Acts". In: *Journal of Child Language*, 2, pp. 1-19.

Chomsky, C., 1968, *The acquisition of syntax in children from 5 to 10*. M.I.T. Press: Cambridge (Mass.).

Chomsky, N., 1965, *Aspects of the theory of syntax*. M.I.T. Press: Cambridge, Mass.

Drubbel, A., 1976, *Enkele begrippen van hoeveelheid by 5-7 jarige kinderen*. University of Amsterdam, Inst. of gen. linguistics. M.A.Thesis.

Ervin, S., 1971, "An overview of theories of grammatical development". In: D.I.Slobin (ed.), *The ontogenesis of grammar*. Academic Press: New York, pp. 189-212.

Felix, S., 1978, *Linguistische Untersuchungen zum natürlichen Zweitsprachenerwerb*. W. Fink: München.

Flanders, N.A., 1970, *Analyzing teaching behavior*. Reading (Mass.).

Geest, van der; see Van der Geest.

Gruber, J., 1967, "Topicalization in child language". In: *Foundations of Language*, 3, pp. 37-65.

Kessel, F., 1970, *The role of syntax in children's comprehension from ages six to twelve*. Chicago.

Koster, K., 1975, *De ontwikkeling van het getalbegrip*. Groningen.

Lawick; see Van Lawick Goodall.

Lenneberg, E., 1967, *The biological foundations of language*. Wiley & Sons: New York.

Levelt, W., 1973, *Grammatical inference*. University of Nijmegen (unpubl.).

MacNamara, J., 1972, "Cognitive basis of language learning in infants". In: *Psychological Review*, 79, pp. 1-13.

Maratsos, M., 1974, "How preschool children understand missing complement subjects". In: *Child Development*, 45, pp. 700-706.

McCawley, J.D., 1968, "Review of the Th.Sebeok (ed.) Current trends in linguistics, III". In: *Language*, 44, pp. 556-592.

McNeill, D., 1966, "Developmental psycholinguistics". In: F.Smith/ G.Miller (eds.), *The genesis of language*. M.I.T.Press: Cambridge (Mass.), pp. 1-103.

Menyuk, P., 1969, *Sentences children use*. M.I.T.Press: Cambridge (Mass.).

Moskovitch, M., 1977, The Development of Lateralization of Language Functions. In: S. Segalowitz/F. Gruber (eds.), *Language development and neurological theory*. Academic Press: New York, pp. 194-212.

Müller-Abé, Y., 1979, *Children's comprehension of idioms;* paper presented at the 'Tagung für Entwicklungspsychologie'. Berlin, October, 1979.

Pechmann, Th., 1979, *Sprache und Gestik als Mittel der Referenz;* paper presented at the 'Tagung für Entwicklungspsychologie'. Berlin, October, 1979.

Rileigh, K.K./Friedlander, B.Z., 1971, *Effect of natural and randomized word order* (unpubl.).

Schlesinger, J.M., 1970, "The grammar of sign language and the problems of language universals". In: Morton, J. (ed.), *Biological and social factors in psycholinguistics*. London, 1971, pp. 98-121.

Schlesinger, J.M., 1971, "Production of utterances and language acquisition". In: D.I.Slobin (ed.), *The ontogenesis of grammar*. Academic Press: New York, pp. 63-101.

Schöler, H., 1979, *Macht der Ton die Musik?* Paper presented at the 'Tagung für Entwicklungspsychologie'. Berlin, October, 1979.

Segalowitz, S./Gruber, F. (eds.), 1977, *Language development and neurological theory*. Academic Press: New York.

Shipley, E./Smith, C./Gleitman, L., 1969, "A study in the acquisition of language." In: *Language*, 45, pp. 322-342.

Slobin, D., 1970, "Universals of grammatical development". In: G.D.Flores d'Arcais/W.Levelt (eds.), *Advances in psycholinguistics*. North Holland Publ.: Amsterdam, pp. 174-186.

Slobin, D.I., 1973, "Cognitive prerequisites for the development of grammar". In: C.A.Ferguson/D.I.Slobin (eds.), *Studies of child language development*. Holt, Rinchart and Winston: New York, pp. 175-276.

Smith, B.O./Ennis, R.H. (eds.), 1964, *Language and concepts in education*. Chicago.

Snow, C., 1972, "Mother's speech to children learning language". In: *Child Development*, 43, pp. 549-565.

Snow, C.E. et al., 1974, *Mothers speech in three social classes*. University of Amsterdam (unpubl.).

Snow, C.E., 1977, "Mothers' speech research: from input to interaction". In: C.Snow/Ch.Ferguson (eds.), *Talking to children*. Cambridge Univ.Press: Cambridge (etc.), pp. 31-49.

Snow, C./Hoefnagel, M., 1978, *Age differences in second language acquisition*. Newbury House Publ.: Rowley (Mass.).

Strohner, H./Nelson, K., 1974, "The young child's development of sentence comprehension". In: *Child Development*, 45, pp. 567-576.

Tervoort, B.Th., 1953, *Structurele analyse van visueel taalgebruik binnen een groep dove kinderen*, (2 volumes). Amsterdam.

Tervoort, B.Th., 1972, "The understanding of passive sentences by normal, deaf, and hard-of-hearing children". In: *Proceedings of the third International Congress of Applied Linguistics, Copenhagen, August, 1972*. Julius Groos: Heidelberg.

Turner, E./Rommetveit, R., 1967, "Experimental manipulation of the production of active and passive voice in children". In: *Language and Speech*, 10, pp. 169-180.

Uden; see Van Uden.

Van der Geest, T., 1972, Some nonverbal aspects in language acquisition. In: *Proceedings of the IVth International Congress on Education of the Deaf*. Stockholm (1970).

Van der Geest, T., et al., 1973, *The child's communicative competence*. Mouton: The Hague.

Van der Geest, T./Snow, C./Drewes, A., 1973, *Developmental aspects of mother child interaction*. University of Amsterdam (unpubl.).

Van der Geest, T., 1974, *Some aspects of communicative competence and their implications for language acquisition*. Van Gorcum: Assen.

Van der Geest, T., 1977, "Some interactional aspects of language acquisition". In: Snow, C./Ferguson, Ch., *Talking to children*. Cambridge University Press: Cambridge, pp. 89-107.

Van der Geest, T., 1978, *Entwicklung der Kommunikation; verbale Interaktion im ersten Schuljahr*. Kamp: Bochum.

Van der Geest, T./Heckhausen, J., 1979, *Sprachliche Anpassung in alters- und erstsprachheterogenen Spielsituationen*. Bochum, unpubl.

Van Lawick Goodall, J., 1971, *In the shadow of man*. New York, etc.

Van Trotzenburg, E.A., 1972, *Ontwikkelingslijnen in het empirisch onderzoek van pedagogische en didactische vraagstukken*. Amsterdam.

Van Uden, A., 1970, *A world of language for deaf children.* Rotterdam.

Wasow, Th., 1973, The innateness hypothesis and grammatical relations. In: *Synthese*, 26, pp. 38-56.

Weir, R., 1962, *Language in the crib.* Mouton: The Hague.

Wittelson, S., 1977, "Early hemisphere specialization and interhemispheric plasticity. In: S. Segalowitz/F. Gruber (eds.), *Language development and neurological theory.* Academic Press: New York, pp. 213-288.

Wode, H., 1978, "Developmental sequences in naturalistic L2 acquisition". In: E. Marcussen-Hatch (ed.), *Second language acquisition.* Newbury House Publ: Rowley (Mass.).

Wode, H./Felix, S., 1979, *Paper on the acquisition of a first and second language.* 'Anglistentagung Berlin'.

NEW WORDS, COMPLEXITY AND ARBITRARINESS*

Florian Coulmas
Research Institute, Tokyo, Japan

Abstract

The creation of novel linguistic items is one of the dynamic forces at work in natural languages. Its significance for a theory of language change is obvious, but its place in a theory of language use is less clear. In this paper we offer some considerations which may contribute to a better understanding of the relationship between historical change and the creative use of language. For this purpose it is necessary to review a number of the established dichotomies and principles of linguistic theory. After briefly deliberating the question of how the conceptual dichotomy between diachrony and synchrony could be psychologically reassessed, a discussion follows about the making of new linguistic signs, concentrating on idioms and compounds. The relationships between complexity and the arbitrariness of linguistic signs thus appear in a new light.

Use and Change

Every so often we are required to say new things not said before. As we encounter new situations, we are forced to make use of our linguistic resources in a way they have never been used. Sometimes we take recourse to a dictionary or explicit inquiry if faced with the task to name an unfamiliar object, but, at other times we simply rely on the linguistic means already available

---

* This work was supported by a grant of the Japan Society for the Promotion of Science, which support I gratefully acknowledge.
I would like to thank John Bisazza for helpful comments on an earlier draft of this paper.

to us. We can always name new things with old words, or old things with new words. Language is an open code and it is in permanent flux.

However, situations new to an individual speaker may not be new to a speech community at large, and hence, the individual in dealing with some lexical emergency may reinvent the wheel and *coin* a new expression which has been common property already but unknown to him. In language acquisition this happens all the time. Only we don't notice it, because a solution to a naming problem that conforms to its conventional solution doesn't strike us as original. Moreover, the conditions of an individual's language acquisition are not usually so well-controlled that it is possible to know whether a new vocabulary item has been coined or reproduced. Yet, children expand their vocabulary not only by taking in new items but also by actively producing items, which is clearly evidenced by the many unconventional solutions to naming problems typical of child language. Usually, such unconventional coinings are abandoned very soon, their use not being reinforced. Children (and adults, too) are liable to expand their individual code by making creative use of its elements. This is one source of language change. But, of course, not every original expression has an effect on the code as a whole. Thus, while language is open to change, individual creativity does not necessarily alter the system. None has ever doubted this fact, but how it is to be appreciated in linguistic theory is not easily agreed upon.

For most linguists a softening of the distinction between synchrony and diachrony in the study of language is anathema. So much have we become used to the artificial and theoretically imposed division between facts pertaining to language 'as it is now', on the one hand, and facts about its historical evolution, on the other hand, that we tend to take it for real, disregarding the highly abstract nature of strictly synchronic analyses.

To point out the precariousness of such analyses is not to question the principle - stipulated by Saussure and adhered to ever since - that diachronic truths and synchronic truths should not be confused. However, given that our aim is to narrow the gap between theoretical notions and empirical facts about language, and given that we want to construct a theory of language as it is used in actual communication, it may be necessary to rethink the way this time-honored dichotomy ought to be applied to the study of how language works.

Consider one of the points Saussure made about the opposition between diachrony and synchrony: "For instance - and to begin with the most apparent fact - they are not of equal importance. Here it is evident that the synchronic viewpoint predominates, for it is the true and only reality to the community of speakers" (1974: 90). Interestingly, Saussure refers to the community of speakers and what is real to them to make his point. The fact that he points out, however, is not quite as obvious as he claims. While the perspective of the ordinary speaker on his or her language is clearly synchronic, it seems doubtful that the notions 'synchronic in a strict sense' and 'psychologically real' can be easily reconciled.

Rather, in as far as it is psychologically real, synchrony is mildly diachronic. "To know to just what extent a thing is a reality, it is necessary and sufficient to determine to what extent it exists in the minds of speakers." This is another quotation from Saussure (from the same page), and it reads surprisingly modern, implying as it does that 'psychologically real' is a graded notion. Synchrony in a strict sense is a notion which is as abstract as that of a geometrical point, that is, an entity having position but no dimensions. In this sense synchrony cannot be ascribed a high degree of psychological reality.

In a more liberal sense synchrony is understood as the state of a language as it is spoken over a considerable period of time, and here it is clearly invaded, or, as some would have it, contaminated by diachrony. Native speakers[1] of a language do have diachronic intuitions extending into the past as well as the future. At any point in time, certain expressions are deemed archaic, obsolete, or old-fashioned, and others as new, current, or original. Judgments of this sort are part of what speakers know about their language. Synchronic as their linguistic knowledge is, it has some diachronic depth. This diachronic depth does not disappear if we concentrate on the linguistic knowledge of a speech community at a given time. It is always there, and it is part of our linguistic competence. Linguistic change is continuous, not discrete. It happens as we use our language, enabling a gradual adaption to permanently changing functional requirements that a language has to meet in order to be the appropriate instrument of communication for its speech community.[2]

In theory, every utterance contributes to change, because no two utterances are alike, and because every speaker partakes in the language. In practice, there are great differences in the extent to which a speaker can exercise an influence on his language. This is most evident for highly standardized languages spoken in mass media-ridden societies. These differences, of course, are due to extralinguistic facts, but it is hard to ignore them, and there can be no doubt that the mass media have become a very important means of channeling linguistic innovation. Politicians, actors, radio announcers and others who have access to the media are the pacemakers, whereas those whose communication range is confined to their family and a couple of friends do not even make the grade of the also-ran. Yet, despite the indisputable power of mass communication, there are limits to the changes it can help to impose on a language. Neologisms must be accepted by the speech community, and while frequent exposure to a new ex-

---

1 For an explication of the notion 'native speaker' see the introduction in Coulmas 1981a.

2 The dialectics of function and purpose and their repercussions on language change are presented very elaborately in Kanngießer 1976.

pression will enhance its acceptability, not every *occasional* neologism is equally likely to become a lexical item of the language in question. Not every innovation is change, because some newly coined expressions fit into the system while others require, or rather, cause minor modifications of the system. Thus, neologisms are not completely random. Like organismic mutations they are subject to constraints by existing structures. If they deviate too much from established structures, they have no chance to enter the system as it evolves.

The systematic constraints limiting the possibilities of innovation are the interlock of diachrony and synchrony, and if they could be captured exhaustively in a system of rules, the conventional distinction[3] between synchronic and diachronic word-formation, that is, between creativity and change in the domain of the lexicon would be completely justified. So far, however, no satisfactory theory of systematic constraints on language evolution has been put forth (with the possible exception of Kanngießer 1976), and the rigid conceptual opposition of synchrony and diachrony in language has been a severe handicap to a better understanding of the interrelations between state and process. This was clearly understood by Coseriu more than two decades ago. Thus a passage from his 1958 book on synchrony and diachrony may serve us as a starting point for further considerations on the expansion of the lexicon.

> Dasjenige, wodurch die Sprache Sprache ist, ist nicht nur ihre Struktur (die nur die Bedingung ihres Funktionierens ist), sondern die Sprechtätigkeit, die die Sprache als Tradition erschafft und erhält. Wenn nun der Wandel als systematisches Werden der Sprache verstanden wird, kann es selbstverständlich keinerlei Widerspruch zwischen "System" und "Wandel" geben, und mehr noch, es darf nicht einmal von "System" *und* "Bewegung" - wie von einander entgegengesetzten Dingen - gesprochen werden, sondern nur von "System *in* Bewegung": die Entwicklung der Sprache ist nicht ein ständiges, willkürliches und ungewisses "Sich-Wandeln", sondern eine ständige *Systematisierung*. ... Mit dem Begriff 'Systemati-

---

3 Cf. e.g. von der Gabelentz 1901, Erben 1975.

> sierung' wird die Antinomie zwischen Diachronie und Synchronie auf grundlegende Weise überwunden (1974: 236).

To capture the essence of *Systematisierung*, and thus to explain every given language as the result of regular processes, has always been the aim of historical linguistics. To Coseriu's credit, it was his intention to vevive this aim which could not withstand structuralist dichotomizing. Yet, however much we sympathize with his case, we have to concede that there is another aspect to linguistic innovation which is complementary to *Systematisierung*. Not every change is systematic. There is a countertendency in language use to blur or give up systematic distinctions, to become vague and underdetermined instead of precise and explicit. Moreover, speakers seem to have considerable tolerance for extragrammatical structures. *Systematisierung* counterbalances indeterminacy and irregularity. If this wasn't so, linguistic change should eventually come to the point where no further systematization is necessary; but such a standstill is an unlikely outcome of historical processes. A more plausible assumption is that perpetual *Systematisierung* is necessary because there is an undercurrent in language evolution undermining or relaxing systematic constraints. In the following we shall discuss some aspects of lexical innovation which can not be subsumed under the tendency of systematization.

Lexical Innovation

Lexical innovation is motivated by a variety of factors, among them the need to name newly invented or discovered objects, species or processes (*astronaut, skyjack*); the desire to replace lexical items which have acquired unwanted connotations or degraded meanings (*intercourse, mistress*); euphemism (*engagement* for *battle*, *mission* for *bombing air raid*); masking (*enhanced radiation warhead* for *neutron bomb*); (pseudo)terminology (*brainwash, supply-side economics*); and the desire to make old things look new (*sleep enhancing system* for *bed*).

The need for new vocabulary items can be satisfied by various means: borrowing, original coining, derivation, compounding, idiomatization. Borrowing is a highly productive strategy which is interesting because in many cases it involves a clash between two phonological systems. Much has been written about the ensuing problems, and we will not go into them here. By 'original coining' I mean the formation of arbitrary signs in accordance with the phonological system of the language. In many languages the morphological forms do not exhaust the possibilities of the phonological system,[4] and in these cases original coining is an obvious option. This option is not very productive, however, because it is difficult to promote such items in the speech community.

Acceptance by the speech community is the chief criterion of lexical innovation, and obviously, a sign which is completely arbitrary is not as easily accepted as one which is not. It is therefore not surprising that derivation and compounding, not being completely arbitrary, are the most productive mechanisms of lexical innovation. The newly invented arbitrary sign is difficult to learn, because nobody knows, or even believes to know what it means, making it difficult to relate the new item to others and thus integrate it into the system. There seems to be a strong tendency to satisfy the permanent demand for new words by reaching out from familiar territory and stretching the meanings of established forms.

Derivation

Suppose, for instance, we needed a verb to designate the action of getting off a space shuttle. Given an appropriate context such as (1), the verb *deshuttle* is immediately understood.

---

4 It could be argued that this is true not for most but for all languages. This view is incorrect because the typical or maximal word length of a language is to be regarded as a property of its morpho-phonological system.

(1) The deshutteling astronauts were welcomed by an enthusiastic crowd of fifteen.

The new verb is made up of familiar elements and in analogy to many other verbs similarly derived from nouns by means of the prefix *de-*, such as *dethrone, derail, deplane, debug,* etc. It is arguable that the base words of these derivational verbs are themselves verbs. This is clearly so in cases such as *debrief, detour,* or *declutch,* but in other cases this possibility has to be ruled out, because the base words are not used as verbs without the prefix. The prefix turns them into verbs, adding some kind of negative meaning. The negation conveyed by the prefix *de-* can be of various different kinds, and the speaker-listener must know the appropriate one. As for *deshuttle,* its analogical links with *deplane* and *detrain* help us to understand the nature of the negation, which does not relate to the activity of shutteling but rather to the state of being abord a space shuttle.

There are two important points to be learned from this example. The first point is that the relationships between the forms and meanings of neologisms are very often underdetermined. This property of lexical innovations will be discussed in greater detail below. The second point is that, while *deshuttle* undoubtedly is a neologism, it is not really a word of the English language. Only very few people have ever engaged in the activity of deshutteling so far, and this situation is not likely to change very much in the foreseeable future. Hence, there is no pressing need to have a word like this. Occasionally, words creep into a language for no obvious reason and without there being any immediate need, but this is not the rule. In most cases a new word is the answer to a specific demand, thus filling a lexical or semantic gap.[5] Theoretically, of course, the vocabulary of every language is full of gaps, but in order for these gaps to be filled they must be felt. Only when an action or object achieves a certain status in the speech community is there any need for a

---

5 Cf. Carroll & Tanenhaus 1975 for one version of the Semantic Gap Hypothesis.

name.[6] It is difficult to make predictions, however, chances for *deshuttle* to become lexicalized are slim.

Returning now to the first point, what does it mean to say that *deshuttle* is semantically underdetermined? The prefix *de-*, as pointed out above, contributes a common element of meaning to a large group of derivative verbs. A possible paraphrase of this quite unspecific element is 'remove or reduce in size or degree'. According to the relationships between this element and the meaning of the respective base word several groups of verbs with the common surface structure de-V can be distinguished. Among these relations are the following:

- a relation between an agent, an object and a quality or its intensity (*desalt, demagnetize, debitter*);
- a relation between an agent, an object and a part of that object or something which sticks to it (*decapitate, deice, delouse, defoliate*);
- a relation between a container and a content and an optional agent (*deflate, deplete*);
- a relation between an agent and a container/vessel (*debark, deplane, detrain*).

The negative element contributed by *de-* also differs in as far as it indicates a gradual denial in some cases (*devalue, decompress*) and an absolute denial in others (*decapitate, deflower*). The verbs of yet another group are somehow negative in meaning, but the negative element is not so specifically bound to the prefix, because it has become an integral part of the word which is no longer detachable (*decline, deceive, decay, decease, defeat*). The prefix itself, however, is still very productive. Many of the verbs derived by prefixing it to nouns, verbs or adjectives are causative (*demobilize, detoxify, decaffeinate*); a much lesser number are agentive (*deplane, defect, depart*); and an even smaller group

---

6 Conversely, the existence of a name suggests the existence of an object, sometimes misleadingly. Upon seeing a word for the first time, the typical question is, "What is phlogiston?" not, "Does it exist?"

comprises verbs with both a causative and an agentive reading *(derail)*.

Theoretically, many more of the verbs of this derivational class could have multiple readings than actually is the case. There is no structural reason why *dethrone* should not be an agentive verb. Just as a plane is something to ride on, a throne is something to sit on. Accordingly, (2) would be a very appropriate sentence for describing the action of stepping down from a throne (literally or metaphorically).

(2) * The king dethroned.

The fact that *dethrone* means 'to remove (someone) from a throne' rather than 'someone) abandons a throne' is presumably due to the historical contingency that kings do not usually abandon their thrones voluntarily.

Conversely, we could think of a context where it is necessary to talk about the removal of airplanes from a surface. Thus, in

(3) * The airport was deplaned before the enemy had a chance to attack.

the relation between *de* and *plane* is the same as in *delouse* where some place is ridded of lice. In (3) the airplanes would be removed from somewhere, while in the standard reading of *deplane* an agent removes him-/herself from an airplane.

Now, as both *dethrone* and *deplane* are established lexical items, these considerations about their possible meanings may appear quite futile. However, similar options exist for the meanings of many of the verbs in this class as well as for its potential members. As another example consider (4).

(4)

	classify[7]	declassify
(a)	to arrange or organize in classes or categories	∅
(b)	to assign a level of secrecy to (government documents)	to remove security classification from

Once data or objects have been arranged in categories, the resulting organization is not usually destroyed, or, if it is, it is replaced by an alternative organization. Thus we have *reclassify* but no reading of *declassify* corresponding to the a-reading of *classify*. This may be plausible, but it is certainly not necessary.

The point at issue here is that *plausibility* plays an important role in determining the meaning of neologisms. Obviously, the requirement that a neologism should lend itself to a plausible interpretation is much weaker than that its interpretation must be unequivocally determined by virtue of the meanings of its parts and the syntactic relations established between them. In many cases context enhances plausibility and helps us to decide on the correct interpretation. *Detoxify* may mean either 'to remove poison from a substance, such as a toxic plant, liquid, etc.' or 'to neutralize or remove poison from an organism that has consumed a toxic substance.' Given an appropriate context, the correct reading suggests itself.

(5) He was awarded a Nobel Price for a drug that detoxifies toadstools.

(5') Gradually she began to feel the detoxifying effect of the milk she had been advised to drink in order to overcome the fish poisoning.

The same holds true for neologisms. They occur within a particular context which is appropriate to their respective intended meaning and, in turn, helps to shape their conventional meaning. Thus, the fact that, despite their being composed of familiar elements, many neologisms are semantically underdetermined if decontextualized,is no problem at all as far as language use is concerned. Of course, the examples examined above are a far cry from being arbitrary signs, but they are semantically underdetermined in the sense that their meanings can not be reliably calculated on the basis of the meanings of their parts alone. This point comes out yet more clearly in the case of many compound words whose correct interpretation hinges to an even greater extent on plausibility reasoning and worldly knowledge.

Compounding

The expansion of the lexicon by means of derivation is a productive and more or less regular process which is generally attributed to morphological processes. In a sense, it is syntax on the word level. Morphology and syntax overlap: Affixes can become syntactically independent words (e.g. *emic, ism)*; syntactic collocations can collapse into morphological units (e.g. *maybe,* French *peut-être,* German *vielleicht*); and elements of compounds can degenerate to the status of suffixes (e.g. German *-mann* as in *Vertrauensmann, Verbindungsmann, Gewährsmann, Landsmann)*. Hence, there is no clear-cut boundary between derivation, compounding, or, for that matter, collocation. That compounds are formed of independent morphemes while derivation involves affixes which cannot occur independently is only a rule of thumb. For most theoretical purposes it is good enough, but it should be born in mind that this rule is often violated by word formation processes. Compounding, in particular, is a borderline area which can be pressed into the domain of either syntax, morphology or word-formation only for the sake of a theory.

Compounds come into being when a problem arises in coping with experientially new situations or with the task of mapping new ideas onto linguistic forms. Not every occasional compound becomes lexicalized, and there are always morphologically complex units for which it cannot be decided whether they are produced by rule application or by drawing from the lexical inventory. The expression of new ideas does not always require new words, but the possibility of combining morphological elements on a sub-syntax level is an important resource of insight and creative thinking. Lexical items function as the anchor ground for the conceptual organization of experience and imagination, a property which is most clearly evidenced by compounds whose integration into the lexicon is proof of a successful and hence preserved solution of a discourse problem. This point needs to be stressed, because (more patently than other domains of linguistic organization) word-formation, and compounding especially, evince the necessity of taking into account the practical dimension of language and its interaction with other systems of the human mind. Unless compounding is understood as a process that secures the appropriateness of language as a tool for communicative and ideational purposes, it will be impossible to explain how compounds are comprehended and integrated into the lexicon.

The fundamental principle of semantics is the Fregean principle of semantic compositionality. This principle stipulates that the syntax and semantics of natural languages are structured in a strictly parallel way. The importance of the Fregean principle is generally recognized, and it is even credited with psychological reality in so far as complex utterances are thought to be interpreted by virtue of the meanings of their constituent parts. The principle of compositionality does not apply to elementary linguistic signs for the obvious reason that they lack internal structure. It should, however, be applicable to all complex signs. But compounds are complex signs. How does, then, the principle apply?

The creation of compound words is governed by word-formation rules of a very general kind which allow for a great deal of freedom.

Thanks to its degenerated inflexional system, English is particularly permissive in this regard. There are hardly any restrictions, for instance, on the formation of nominal compounds in English. As long as it makes at least some kind of sense almost any two nouns can be juxtaposed to form a temporary or permanent unit. But what are the semantic criteria for the juxtaposition of nouns?

One rule is that the modifying element must precede the modified element. Thus we have *bookshelf, case grammar, housework, workhouse,* etc. Housework is a particular kind of work, and workhouse is a particular kind of house. This principle of 'modifying-before-modified' (MBM-principle) is so general that it sometimes imposes upon conceptualization. The Random House Dictionary defines *waterfall* as "a steep fall of water from a height", yielding to the general principle that suggests that a waterfall is a particular kind of fall. "Water that falls from a height" would at least be as adequate an explanation of the word *waterfall*, in which case *fall* would be conceived as modifying *water*, and not vice versa. Concept-formation and word-formation intertwine, and the success of a nonce-compound very much depends on its ability to evoke a conceptual unit. Considering the diversity of the relations subsumed under the general heading of modification, and considering the underdeterminacy of many compounds, it is not very likely that this ability to appeal to the fancy of the listener relies to a large extent on the unconditional applicability of the compositionality principle. In view of the overall goal of integrating word-formation processes into a theory of language use, this point is important and worthy of further elaboration by way of illustration.

Consider the large group of nominal compounds whose first element is *air*. Some examples are *air bag, air base, air brake, air cortain, air drop, air foil, air frame, air gun, air lane, air lock, airman, air piracy, air pocket, air pollution, air raid, air space, air superiority.*[8]

---

8 All of these examples as well as the entries referred to below are quoted from The Random House Dictionary, 1978.

As would be predicted by the MBM-principle, *air* is invariably the modifying element in these compounds - which is about all they have in common. There is just one syntactic device, i.e., the juxtaposition of two nouns, to cover an apparently unlimited variety of semantic relations. No two of the listed examples can be assigned the same underlying case relations.[9] Moreover, they are, without exception, to a greater or lesser extent underdetermined and, accordingly, they differ with respect to immediate understandability, or, self-explanatoriness. A common relation in nominal compounds is the relation between a thing (modified) and the material it is made of (modifying). Air is not normally the stuff that things are made of. Yet, if the need arises, a metaphor can be readily coined and air becomes the material of which an object is made, as in *air curtain*. Presumably, an air space is a space made of air, but this would be a very inadequate definition of the word *air space* which is a technical term in the language of international law denoting "the region of the atmosphere above a nation over which it has jurisdiction." Everyone who knows what air is and what a gun is will rule out the possibility that *air gun* denotes a gun made of air. But there is no way of infering the particular nature of the form-meaning link of the word *air gun* from our knowledge of the meanings of *air* and *gun*. An air gun could be any of a number of things, such as a gun for shooting in the air, a gun for shooting from the air, or a gun operated by compressed air. Similarly, *air bag* could denote all kinds of things having to do in some way or other with air and bags, among them vomit bags, light suitcases especially designed for air travel, and a safety device "that inflates automatically upon impact in order to protect automobile passengers from injury in a collision." In the latter case air is the content of the bag, but in the former two it is not -- that is, not primarily -- and the respective relationships between *air* and *bag* in the former two cases are more remote.

Notice the superficial similarity between *air bag* and *air pocket*. It is so conspicuous that it is difficult to present a convincing argument why an air pocket could not be called *air bag*. One

9 Cf. Fillmore 1968.

possible rationale is that part of the meaning of *pocket* is that a pocket is attached to something. An *air pocket* is metaphorically attached to, or is part of, the sky, and thus is something into which an airplane can fall. An air pocket, in a manner of speaking, is void of air. Except for the common element *air*, there is no link whatsoever between *air bag* and *air pocket*. There are, of course, nominal compounds with *air* whose underlying case relations are the same. Thus we have items such as

(6) *aircraft, airplane, air train, air taxi*

and

(7) *airport, airfield, airstrip.*

The compounds in (7) are contractions. An airstrip is a runway for *airplanes*, and *air field* denotes "a level area on which airplanes take off and land." Again, the contractions entering into compounds can be of various kinds. In one of its meanings *air bag* denotes an object that you find on the airplane in the seat pocket infront of you. Here *air bag* is short for *airsickness bag*, and *airsickness* is short for *a feeling of nausea induced by the movements, changes of altitude, air pressure, etc. of flying airplanes*. An air brake is a device for reducing the speed of an aircraft or, alternatively, a brake operated by compressed air. In either case the surface representation of the respective meaning is very incomplete. A lot more than knowledge of the meanings of the compounded words is called for in order to understand the compounds.

Is polysemy a theoretical remedy for this practical flaw? Well, no, and there are two very good reasons why it is not. First of all, this kind of underdeterminacy is no flaw. We can live with shortcut compounds, no matter how much they upset a theory which is based on the compositionality principle of linguistic meaning. Most of the time, we don't even notice that they are contractions. There is no desire nor need to be overexplicit. Second, the assumption that *air* is polysemous having different meanings, such as 'air sickness', 'aircraft', 'compressed air', etc. and that a selection of the appropriate meaning is made according to context, is ad hoc. Although *air* is semantically ambiguous and vague, it never has any of *these* meanings in isolation. In some cases the interpretation of compounds can be described as the mutual selection of one of several meanings of the constituents. Thus both *court* and *yard* have several meanings, but *courtyard* is unambiguous, because *court* and *yard* have one meaning in common which is mutually selected by juxtaposing the two elements. But this procedure cannot explain contracted compounds.

Contraction is one way of loosening the one-to-one relation between forms and meanings. It is a highly economical strategy for enlargening the vocabulary, as it allows the recycling of morphological elements for many different purposes. The price to be paid for this possibility is underdeterminancy. In contracted compounds, as well as in many others, a part of the meaning of the compound as a whole has no formal correlate and therefore has to be supplied by the speaker. Without having to analyze the above examples further, it is clear that compounds differ widely in terms of their self-explanatoriness. The speaker has to supply a small part of the meaning of the whole in some cases and a bigger part in others. However, a very large portion of the vocabulary consists of compounds, and, as a rule, their interpretation poses no particular problem. The task of compensating for their underdeterminacy is accomplished via a number of principles that make up for the shortcomings of the compositionality principle. Whenever it fails, they can take over. Most important among these principles are ANALOGY, METAPHOR, CONTEXTUAL INFERENCE, WORDLY KNOWLEDGE, and IMPLICATURE.

Although there is no obvious limit to the diversity of the underlying relations obtaining between the parts of compounds, many compounds are modelled on others. They frequently belong to analogy chains[10], some of them being members of more than one such analogy chain. *Air brake*, for instance, stands in an analogy relation with *air gun* in one of its readings ('operated

---

10 The best account of analogy to date is Anttilas 1977 overview.

by compressed air'), and with *air frame* in another reading ('a part of an aircraft'). Part of the job of giving an interpretation to a compound, or of producing one, is to accomodate it in the intricate network of analogy chains that is part of the organization of the lexicon. Analogy is a relation not of sameness but of similarity. Thus, coining a compound by analogy to another more often than not involves an expansion of existing possibilities. On the other hand, interpreting a compound by analogy means to rely, partly at least, on familiar structures.

Metaphor is a kind of analogy. It involves reasoning by analogy and is indispensable for fitting new experiences to our previous knowledge and beliefs. Metaphor enables us to make use of available forms in the most economic way: Old words are constantly equipped with new meanings by being pressed into new uses by analogy. Compounding is a very fertile ground for metaphor, precisely because so many compounds are underdetermined and leave so much of their interpretation to the speaker. If a sensible interpretation is not achieved on the basis of the literal meanings of its parts, the best guess is that a compound is a metaphor. However, this account suggests that the metaphorical reading is secondary to its literal counterpart. For many cases this seems quite unlikely. The most ordinary words are metaphors or metaphoric in origin, and some metaphors are so pervasive that they are hardly recognized as such.[11] Once metaphors such as *airport*, *air pocket*, *airwaves*, etc. have become conventionalized they behave just like regular names. We are reminded of their metaphorical origin by more recent neologisms, such as, *air train*, *air taxi*, for which they served as analogical models.

Metaphoric or straight, in many compounds the conjoined forms are not enough to yield unequivocal interpretations. New words ought to be comprehensible to a native speaker without instruction.

---

11 It is arguable therefore that metaphor in the strict sense is a process rather than a property of expressions. It is a mechanism of creating new relations between forms and meanings. Successful metaphors often turn into idioms. Cf. Coulmas 1981b: 6f.

This is a major precondition for their easy spreading. Yet, as pointed out above, there is no need to be overexplicit. Rather, a principle of least effort and greatest effect seems to apply in order to facilitate effective communication. The implementation of this principle is aided by contextual inference. A compound such as *air lock* which, taken by itself, is underdetermined, is much more likely to be understood correctly if encountered in a context such as (8).

(8) The pressure chamber can only be reached through the air lock.

In many cases the interpretation of a compound is to be understood as a function that relates a pair consisting of a compound and a context with a meaning.

Contextual inference usually interacts with worldly knowledge and implicature. *Air pollution* and *mercury pollution* exhibit identical surface structure, yet they are members of two different analogical chains, the former being pollution *of* the air and the latter pollution *by* mercury. The fact that mercury is potentially dangerous is part of our worldly knowledge, and therefore 'polluted percury' is a very unlikely reading of *mercury pollution*. Air, by contrast, is generally considered to be much more liable to be polluted than to pollute.

Accordingly, air-conditioning is conditioning of the air, and not by air. Notice, incidentally, that the noun is *air-conditioning*, whereas *air-condition* only exists as a verb. *Air-condition* would make a straightforward noun designating, for instance, the relative humidity, air pressure, etc. in the context of meteorology.

(9) The prediction for today's air-condition is ...

Compositionally this would be a reasonable interpretation of the compound *air-condition*. That this possibility will ever be made use of, however, is very unlikely owing to the prior employment of the form *air-condition* as a transitive verb in a different context.

From our discussion so far it seems clear that compounding, as one of the most prolific word-formation strategies rests on the integration of different psychic systems and skills, linguistic as well as cognitive and perceptual. The systematic place of compounding is between speech and language, because the rules that allow us to form neologisms clearly belong to language, while many occasional neologisms never make it into the lexicon. Others however do, and in conclusion we will briefly consider some aspects of lexicalization.

Lexicalization

"If a linguistic form exhibits internal structure, it can be synthesized and need not be stored in the lexicon, and if a form needs to be stored in the lexicon its meaning cannot be synthesized." The observations presented in this paper have revealed some important exceptions to this general principle of "either not complex or synthesizable." Psychologically speaking, it is clear that the lexicon is highly redundant, containing many complex expressions. However, there is still a tendency for the above principle to be reinforced or restored. It can be seen at work in compounds and idioms whose internal structure is fading.

Generally, loss of internal structure is a process that assists a construction in becoming conventional and part of the lexicon. Hundreds of words are former compounds or derivatives. Their parts have become so firmly glued together that their individuality is all but lost. In some cases the juncture is still recognizable but insignificant, in others it has disappeared. If two elements are repeatedly juxtaposed, and the juxtaposition develops a life of its own, it becomes lexicalized. In the process of lexicalization the compounded parts typically lose their semantic independence before this is reflected by the collapse of the formal structure. What follows inevitably is a break-up of the parallelism of form and meaning. As a result there are always complex forms with simplex meanings. The effect that such an unstable situation has on the speaker-listener was lucidly described by Sapir, surely a linguistic dynamist:

> Where there is uncertainty about the juncture, where the affixed element cannot rightly claim to possess its full share of significance, the unity of the complete word is more strongly emphasized. The mind must rest on something. If it cannot linger on the constituent elements, it hastens all the more eagerly to the acceptance of the word as a whole.[12]

When a compound is about to become lexically viable, there must be a time when speakers are undecided and uncertain of whether the parts or the whole should be foregrounded. This may be more obvious in morphologically rich languages than in isolating languages. German is a case in point. A list of adjectival examples follows.

(10) German compound adjectives with competing superlative forms

Positive	Superlative internal	Superlative external
a. naheliegend	nächstliegend	naheliegendst
b. schwerwiegend	schwerstwiegend	schwerwiegendst
c. weitreichend	?weitestreichend	weitreichendst
d. hochgelegen	höchstgelegen	?hochgelegendst
e. nahestehend	nächststehend	nahestehendst
f. vielgelesen	meistgelesen	Ø
g. zartfühlend	Ø	zartfühlendst
h. hochtrabend	Ø	hochtrabendst

These compounds are composed of adjectives (functioning as adverbs) and present participles. The MBM-principles holds in as far as they are not considered as unitary wholes. Thus, the adverbs modify the participles, just as they would do in a syntactic construction. In all of the examples in (10), and in many other such

12 Sapir 1921: 132.

cases, frequent juxtaposition has led to lexicalization, although the compounded parts are clearly recognizable. The salience of their identity, however, varies. As long as morphological rules operate on the individual parts they preserve partial autonomy. If the meaning of these compounds is interpreted compositionally the more reasonable formation of comparative and superlative forms operates internally, because it is the degree that varies, and not the property, the former being expressed by the initial members of the compounds. The external formation, by contrast, indicates an increased cohesion between the compound parts whose conceptual salience is reduced.

The comparative or superlative can be formed from the first members of the compounds in isolation, but not from the second parts. Nevertheless, as exemplified by (10g and h), there are cases where the internal option is no longer applicable. **Höchsttrabend* or **höhertrabend* are unacceptable, because the fusion of the elements is too advanced to allow their separation. This is due partly to the metaphoric origin of *hochtrabend* ('pompous'). By forming the comparative or superlative internally, the individual parts are conceptually foregrounded, inviting a compositional interpretation. However, the normal speaker is no longer aware of the metaphorical trace of *hochtrabend* and is not prepared to reconstruct it by first giving literal interpretations to the parts. As horseback riding (the domain from which the metaphor stems) is no longer part of public life, the literal reading of *hochtrabend* is all but incomprehensible to most speakers of German. The metaphor is no longer considered as such. *Hochtrabend* has become idiomatized, the independence of its parts is lost, and morphological rules can only operate on the word as a whole.

The impossibility of treating the formally discernible constituents of *hochtrabend* as partly autonomous elements is a result of the bleaching out of its underlying metaphor. This hypothesis is supported by further evidence:

(11)	a. großspurig	*größerspurig	großspuriger
	b. hochgestochen	*höhergestochen	hochgestochener
	c. kleinkariert	*kleinerkariert	kleinkarierter
	d. vielversprechend	*mehrversprechend	vielversprechender

Compare (10f) and (11d). The former has only a literal reading, the only expected reading of the latter is metaphorical. But as is the case with (11a-c), the metaphor has become conventionalized to the point that a semantic interpretation of the elements on a one-by-one basis no longer yields a plausible result. *Groß, klein, hoch* and *viel* are among the most general and widely used grading adjectives, yet they have succumbed to the unity of the compounds which have turned into idiomatic lexemes.

(10a, b) and (e) show that the uncertainty about where to apply the morphological rule has grown into a distinction. Both *nächststehend* and *nahestehendst* are acceptable, but speakers report subtle differences in meaning associated with the two coexisting types of forms, which again backs the above hypothesis. *Nahestehend* is ambiguous, having a literal (l-)reading of spatial nearness and a metaphorical (m-)reading of emotional closeness, the latter being strongly predominant.[13] This ambiguity is only partly inherited by the comparative and superlative forms. *Nahestehendst* (external) is much more likely to be associated with the holistic m-reading and, conversely, the compositional l-reading is more likely for *nächststehend* (internal). The morphemic atom is focussed by internal marking, whereas the compounded molecule is foregrounded by external marking. It should come as no surprise that a differentiation in form attracts semantic differentiation in turn. In cases such as (10e) this process involves an interesting shift from the synthetic to the holistic in the form-meaning relation of compounds.

A last question in this connection concerns a third possibility

---

13 The fact that the idiomatic or metaphorical readings of idioms are far more frequent than their literal counterparts has often been observed; cf. e.g. Chafe 1970.

for forming the comparative and superlative forms of compounds that combines the other two possibilities discussed above. Some examples follow.

(12) a. der näch*st*liegend*st*e Gedanke
b. die *meist*gelesen*st*e Zeitung
c. grö*ßt*möglich*st*e Eile
d. die *best*angezogen*st*e Frau

(Bold print indicates superlative marking.)

How is the meaning-form relation to be understood in these cases? According to German reference grammars and textbooks, constructions of this tape are simply wrong and have no right to exist, because the superlative is redundantly marked twice. Yet, they do exist; as a matter of fact, they are widely used. Redundancy is, of course, no absolute reason for a form to cease to exist, and furthermore it is very much the question whether the formal redundancy is really considered redundant by the speakers using these forms. (12c) is an obvious case. *Größtmöglich* is a lexicalized superlative lacking a comparative form. As it is not part of a paradigm, it seems quite likely that the holistic perception of the word has become so dominant that the internal marking of superlative has lost its force. Again a formal distinction has dissociated itself from a corresponding distinction in meaning, preparatory to being recycled in the process of never-ending innovation that depends on the possibility of independent changes in form and meaning. Inevitably, this process upsets the parallelism of form and meaning. It is counter to the tendency of *Systematisierung* (cf.pp.231f above) and sets the limits to the compositional interpretation of complex forms.

At any given time, all complex expressions range along a continuum from fully determined to arbitrary. At one end of the continuum, where compositionality holds, the rule of syntax is unchallenged. At the other end, where forms lose their semantic independence, the form-meaning relation of complex forms is characterized by underdeterminacy and eventually arbitrariness.

Arbitrary complex linguistic signs are not supposed to exist, but they do.

Bibliography

Anttila, R., 1977, *Analogy*. Mouton: The Hague.

Carroll, J. M./Tanenhaus, M.K., 1975, "Prolegomena to a Functional Theory of Word Formation". In: *Papers from the Parasession on Functionalism*, M. E. Grossman, J.L. San, T.J. Vance (eds.), Chicago Linguistic Society, 1975, 1, pp. 47-62.

Chafe, W., 1970, *Meaning and the Structure of Language*. University of Chicago Press: Chicago, Ill.

Coulmas, F. (ed.), 1981a, *A Festschrift for Native Speaker*. Mouton: The Hague.

Coulmas, F., 1981b, "Conversational Routine". In: F. Coulmas (ed.), *Conversational Routine. Explorations in Standardized Communication Situations and Prepatterned Speech*. Mouton: The Hague, pp. 1-17.

Erben, J., 1975, *Einführung in die deutsche Wortbildungslehre*. Erich Schmidt: Berlin.

Fillmore, C.J., 1968, "The case for case". In: E. Bach/R.T. Harms (eds.), *Universals in Linguistic Theory*. Holt, Rinehart and Winston: New York, pp. 1-88.

von der Gabelentz, G., 1901, *Die Sprachwissenschaft. Ihre Aufgaben, Methoden und bisherigen Ergebnisse*. Leipzig.

Kanngießer, S., 1976, "Sprachliche Universalien und diachrone Prozesse". In: K.-O. Apel (ed.), *Sprachpragmatik und Philosophie*. Suhrkamp Verlag: Frankfurt, pp. 273-393.

Sapir, E., 1921, *Language*. Harcourt: New York.

Saussure, F. de, 1974, *Course in General Linguistics*, New York.

SPEECH ACTS AND CONTEXT CHANGE

Marina Sbisà

1. In this paper I would like to argue for the view, that the primary concern of a speech act theory ought to be the study of the changes that speech acts bring about in their own contexts.

I do not hope to be wholly convincing. Any exhaustive argumentation on this topic would involve a thorough discussion of general methodological and philosophical issues. Moreover, I feel that some assumptions of mine clash with other assumptions that are widely shared by students in pragmatics. However, this will be shown rather than said by this paper.

What I would like to declare from the very beginning is that the notion of speech act I am concerned with is Austin's one. Nearly twenty-five years after *How to Do Things with Words* was published posthumously (but more than thirty years after Austin first delivered his Oxford lectures on *Words and Deeds)* the true philosophical import of his work is becoming clearer and clearer, in spite of all analytical, post-analytical and anti-analytical fashions.

Searle's speech act theory was inspired by Austin's work, but it should not be mistaken for a realization of the austinian project. After all, to gather up all the suggestions put forward by Austin would have been a hopeless enterprise: some of them conflict with each other, as is the case (to quote a crucial example) with the 'locutionary act'. Sometimes Austin seems to consider the locutionary act as that part of the speech act, which is deemed true or false (Austin (1975, p.148)); elsewhere, he seems to consider it as a necessary component of a speech act, in order for the latter to be deemed true or false *as a whole* (Austin (1975, p. 145)). He also includes referring into the locutionary act, as being "ancillary" to the rhetic act (Austin (1975, p. 95)), while elsewhere he deals with reference as a widespread felicity condition of illocutionary acts (Austin (1975, p. 50-52)). Facing such shortcomings, Searle (1969) redefined the locutionary act as a 'propositional' act, providing once again a stable stand-in for what may be called true or false. This redefinition made the theory smoother and more workable, but had inevitable consequences as far as the content of the theory and its relevance were concerned. Austin claimed that statements are acts, and so does Searle. But according to Austin, in order to understand the relationship between assertive and non-assertive speech acts we ought to investigate a family of related judgements on speech acts in situation (like 'fair/unfair', 'sound/unsound', 'right/wrong', and even 'good/bad': cf. Austin (1975, pp. 141 ff.)); whereas Searle (1975) accounts for the same problem by distinguishing among different 'directions of fit' between the propositional content of the speech act and the 'world'. Of course, I do not want to argue here about who is right and who is wrong; what I want to do is merely stress the fact that such differing answers to the same question involve different notions of truth as well as different definitions of the statement. It seems to me that Austin's tentative view of truth and falsity is interestingly *dynamic* (whereas Searle's directions of fit connect two *static* terms with each other), and that it is appropriate for the implicit, ambitious project to hit the theory/praxis dichotomy in its very heart (a linguistic one: the privileged status of 'constatives'). It is within such a framework that the claim that statements are acts can be relevant (though, of course, questionable). It is also because of the fact that this framework has been put aside, that for the statement to be called an 'act' has become little more than an unquestionable, and therefore uninteresting, label.

Other differences between Austin and Searle are perhaps better known. Austin was no wittgensteinian, but his work is curiously respectful of the saying/showing distinction (the locutionary act is an act of saying, while the illocutionary act is shown by force-indicating devices ...). If on the contrary, according to

Searle's expressibility principle, what can be meant can be said (cf. Searle (1969, pp. 19-21)), such a distinction collapses and what is left is a mono-dimensional language where both meanings and forces exist in so far as they can be explicitly said, and where resorting to deep structures and/or conversational inferences can hardly account for non-explicit signification. More importantly for the topic of the present paper, it may be noted that the concept of an 'act' for Searle is a positive one, whereas Austin defines the various sub-acts within the total speech act in a negative way (they can be distinguished from each other only "by means of the possible slips between cup and lip": Austin (1975, p. 147)). Searle wants to define speech act concepts by listing all their felicity conditions; Austin, following Hart (1951), is interested in those features that make speech acts 'defeasible'.

Of course, Austin's work does not present us with a perfectly elaborated speech act theory. But, on the whole, we still find a great deal of unexploited indications, and even some conceptual prerequisites for a speech act theory to be philosophically worthwhile. I assume that working out such indications can help serve the purpose of gaining a better (and perhaps dynamic) understanding of linguistic action.

2. Act, Action and Interaction

In order to talk of speech acts (and in particular of illocutionary acts) as a kind of action, we shall adopt a number of more or less technical definitions. First of all, it is important to remember that 'act' and 'action' are not synonyms.

An 'act' is a unit to be isolated on the background of an agent's activity. 'Activity' itself involves reference to some 'active' (as contrasted to 'passive') attitude or behaviour, occurring throughout a certain period of time. Unlike Von Wright (1968, p. 39), I do not think that the notion of 'activity' is to be related to the notion of process, since an activity lacks internal temporal orientation: if any temporal unrolling is taken into account, it is viewed as cyclic. An activity could rather count as a state, to be characterized by the constant presence, or by the cyclic repetition, of some 'active' behaviour (cf. also Beardsley 1978, p. 167). Whereas to describe an action usually involves mention of goals and purposes, in order to describe an activity reference to goals and purposes may be dispensed with.

'Action' refers to a process. As Shwayder puts it, "'Action' directs our attention to what happens, to the movements made and the results thereof" (Shwayder (1965, p. 31). Also Ware (1973) stresses this feature of 'action', by claiming that we can watch actions (by watching physical movements and the changes they bring about), but not acts. But are 'action' and 'act' two separated, mutually exclusive kinds? I am tempted to suggest that their difference does not lie in the phenomena referred to, but in the perspective (as it were, the 'aspect': cf. for example Comrie 1976) in which these phenomena are viewed. 'Act' involves a perfective aspect: a doing is considered as a whole, without taking into account its internal temporal articulation.By the way, this could also account for the idea that acts are the units by which activities are constituted. 'Action' involves an imperfective aspect: that is, attention for phases, results, means-to-ends relationships, and the like. Thus, acts are not always, but sometimes can be, considered as actions; and actions are not always, but sometimes can be, considered as equivalent to acts or to collections of acts. 'Act' and 'action', in this perspective, seem however to have a common core: the achieved result.

'Interaction' and 'action' should also be distinguished carefully. Of course, not all action is mutual action. But, as I want to argue below, if illocutionary acts are actions at all, they belong to an interactional sub-type. Moreover, there are reasons for giving to 'action' what may be called an 'interactional' definition.

2.1 In discussing actions (but also acts and activities) we are obliged to deal with the ways in which we talk about them. We must take into account ways of reporting, narrating, and describing actions, that is, ways of giving representations of actions by linguistic means. Of course, as is argued by Davidson (1980, p. 166), one may, and in fact one does refer to events in ordinary talk; the same may be said of actions, even though not all actions are events (I agree with Brennenstuhl (1975) on this point). In my opinion, it is questionable whether our ordinary ways of referring to events and actions commit us to an ontology of such entities; but this will not be discussed here. I merely want to claim that it is very difficult (if not completely impossible) to identify an action in a merely *deictic* way. Compare:

(1) Look there!

(2) Do so and so ...

(3) Take this.

(1) drives the hearer's attention to something that is happening (perhaps, an action), but cannot be said to identify an action or even an event: the deixis at work ("there") is merely spatial. (2) can be understood by the hearer as a proposal to perform a certain action, but needs accompanying gestures of an imitative kind; moreover, does "so and so" refer to these gestures, to the action to be performed, or to both? The situation is unclear but, certainly, not merely "deictic". In (3), however, apart from all questions about how exactly the hearer comes to understand the intended reference, "this" does deictically refer to an object, for example to the coin in the speaker's right hand. This does not simply amount to saying that actions are usually referred to by way of descriptions: such a claim would be trivial, since the same happens to objects. I want to suggest that whenever an action is referred to not by a description, but (say) by a pronoun, it is anaphora, or cataphora that occur, and not deixis. The action to be picked out can be identified only by reference to previous (or subsequent) reports and descriptions, or to non-linguistic equivalents of these, such as 'keyed' occurrences of actions (cf. Goffman 1974, p. 45, p. 58 ff.)) belonging to the relevant action

type.

One might add that while one may catch an unknown beast before having a name for it at one's disposal, one cannot 'catch' an action, or (more seriously) single it out, unless by linguistic, or by other semiotic means. With respect to this, human cultures seem to function rather like huge storages of narrative schemata, according to which people understand (describe, report, narrate, evaluate, explain ... and identify) each others' actions. This is surely why the analysis of narratives was able to give an important stimulus to semiotic studies, and did not fail to provide, at later stages, useful suggestions for the semiotics of discourse (cf. Greimas (1966), (1970), and (1983); Greimas and Courtés (1979)). For the same reasons, ethnomethodology can take as its starting point that "concern for the nature, production, and recognition of reasonable, realistic, and analyzable action", which operates in most everyday situations quite apart from the participants' interests in philosophy or in professional sociology (Garfinkel (1967, p. 75)). As a matter of fact, a large amount of our everyday talk is devoted to reporting (or commenting on) what someone did, what others did to him/her, what we (or you) have been doing, what we (or you) are going to do. But, in connection to such matters as the interactional definition of social situations, relationships, and roles, there is no point in asking for what 'really' happened apart from the participants' own understanding of each others' actions and from the descriptions or reports they do (or would) agree upon. It is considerations like this which indicate the suitableness of a connection between the definition of 'action' and the phenomenon of something being taken as an action by an observer/participant.

2.2. With respect to the aims of the present paper, there is no point in giving to 'action' a causal definition. To invoke intentions and various pro-attitudes as causal factors distinguishing actions from events (as is done, for example, by such authors as Goldman and Davidson) leaves unexplained at least as much as it explains. What it amounts to is re-introducing the old philosophical subject and its metaphysic point of view: who else is to deem whether there is a causal connection between what? Moreover, in actual interaction intentions, wants, desires, and the like are usually recognized from the mode of the action, and such a recognition cannot but presuppose consideration of a certain stretch of behaviour as constituting or trying to constitute an action. Even if we should concede to Davidson that there can be something like a "pure" intending, separated from the intended action or any attempt at it (Davidson 1980, pp.83ff.), its existence does not demonstrate that the analysis of action depends on the analysis of intention.

Here, we shall propose an interactional definition of 'action', whose purpose is not to pick out the set of properties (if there are any) that make something 'really' an action, but to describe what happens when something is viewed as an action. I would like to distinguish the following steps:

(1) In a situation S, some participant P_1 picks out a state St_1;

(2) P_2 contrasts St_2 with a presupposed state St_1;

(3) P_2 considers St_2 as brought about (by way of contrast to St_1) by an agent P_1, or under P_1's responsibility.

Schematically:

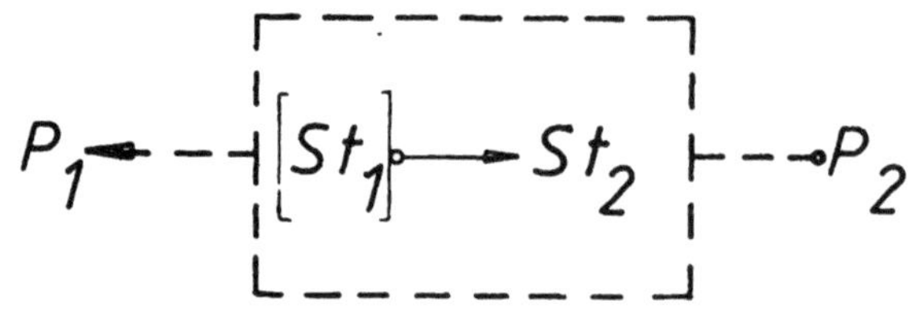

Fig. 1

Some brief comments about such a proposal are due.

Here 'action' is dealt with as an 'ascribed' concept. I take it that such an ascription functions along the lines expounded in Hart (1951). That is, I am willing to allow for the 'defeasibility' of action. By this I mean that, in order to ascribe a certain change-in-the-world to the responsibility of an agent, we do not need to display ultimate evidence; but, whenever the ascription raises doubts, it will be submitted to discussion. It is always possible to discover new facts about the alleged action, and/or to reframe its description, so that P_1 can no longer be held responsible for it. In certain cases it is even possible to switch to a merely "natural" framework, instead of the previously adopted "social" one (cf. Goffman (1974, pp. 21-22)). The introduction of *degrees* of defeasibility and/or of defeasance could perhaps account for puzzling cases such as those in which the agent did something within his/her control in an uncontrolled way (cf. Brennenstuhl (1975, pp. 83 ff., and pp. 209 ff.)).

It could be objected that there are such 'simple' (or perhaps 'basic') actions as raising one's arm, whose description as actions does not involve any decision like those decisions typical of ascriptions (if these are opposed to descriptions: cf. Feinberg (1964)). But even in such cases, if we admit with most theorists that such simple actions do not count as actions unless they are intentional, but we also recognize that it is the observation of behaviour that makes intentions recognizable, it turns out that the behaviour that counts as the performing of an action is itself the grounds for us to interpret it as an action. In considering that behaviour as constituting an action there is, therefore, an unavoidable element of decision.

Our P_1 can be a human individual, an infra-individual instance (e.g. the Unconscious), a super-individual construction (class, party, society, church, and the like), and even a personified natural agent. 'The action of wind' means, indeed, that were it not for the wind things would have gone otherwise, even though the wind is envisaged within a natural framework and therefore cannot be said to have intentions or to control the changes it brings about.

In this connection the problem (tackled by causal theories of action) of whether and how the notion of action is linked to the notion of cause crops up again. But here one might feel tempted to say, as is done (how seriously?) by Austin (1970, p. 202), that our idea of 'action' is primary to that of 'cause', and that the latter arises whenever ascription of responsibility is switched down to a "natural" framework. (This might be one more reason to dispense with 'events' when discussing action).

Contrary to P_1, our P_2 is always human: an individual (possibly, the speaker talking about someone's actions) or a group of individuals (including, perhaps, the philosopher who is discussing action theory).

As for St_1, it should not be identified with the actual initial state, since the latter could even be unknown to P_2. However, the words used to identify an action usually involve reference to the initial state relative to that action, and therefore to St_1, whether this corresponds to the actual initial state or not, so that a St_1 is *presupposed*. It is not clear to me whether the initial state of an action can ever be described apart from the description under which the action itself is identified.

As for St_2, it is neither the final state of the action process in a merely temporal sense, nor any kind of effect or consequence, but the result of the action, that is, that which has to be the case if the corresponding action is to be a successfully performed one.

I have omitted reference to counterfactual states of affairs (cf. Von Wright (1968, pp. 43-45)). Of course I agree that typically, without the action, things would have gone otherwise. But in the present case this does not need to be explicitly stated. Our P_2

does not always need to refer to a definite state St_3, that is, what would have been the case, had it not been for the agent's interference. In particular, in situations in which every stretch of behaviour is liable to be understood as an action (as it is usually the case in face-to-face interaction), the counterfactual state St_3 ought to be replaced by any one of the results St_2', St_2'', St_2''', and so on, of any utterance, gesture or silence on the part of P_2, other than the one he/she actually produced.

As for the agent's intentions, I do not explicitly include them in the representation of the action, but I take it that by identifying P_1 as the agent, P_2 (together with us) considers P_1 as endowed with the necessary 'competence' for the ascribed performance, including, where suitable, the intention to bring about St_2 by certain means. Of course, an intention can also be expressed by behaviour different from the successful performance of the intended action, and can thus be attributed to a subject who tried but did not succeed in performing the intended action, or has not even tried to perform it yet. Such complications however, which involve the identification of an intended but not realized St_2, do not constitute real counterexamples to our proposed analysis of 'action'.

2.3. The proposed perspective could imply some consequences which I am not willing to accept. It might seem that if something's being an action depends on its being described as an action, it depends also on the availability of a certain action verb in the lexicon of a natural language.

Austin seems to accept this view, since, in order to investigate linguistic action, he turns to linguistic action verbs. Even more radically, recent research in the speech act verb lexicon (cf. Ballmer and Brennenstuhl (1981)) relies on the assumption that the organization of such a lexicon corresponds to the organization of the field of linguistic action. In my opinion, we should not be too hasty in accepting assumptions of this kind.

Where the lexicon of a natural language draws a distinction of any kind, language theory ought to maintain and possibly to exploit it. But when a certain distinction is not explicitly present, this cannot be taken to mean that there is no point in drawing it. A distinction can be quite inapparent but nevertheless at work; it may be useless in practical contexts where nobody and nothing challenge it, but useful whenever we are in danger of introducing confusions into a would-be conceptual clarification. I am thinking, for example, of the distinction between illocutionary and perlocutionary acts.

Moreover, in actual interaction, metacommunication is carried out mainly by implicit means: but this does not mean that the participants fail to understand each other as performing communicative actions. Of course, the participants may give explicit definitions to what is going on; but whenever this is done, the descriptions turn out either to be generic and ambiguous enough to allow for further play, or else to be quite selective with respect to the implicit indications they are claimed to account for. Even in non-metacommunicative reports referring to previous encounters or to participants other than the presently involved ones, the most precise way of telling "what he/she really did by/in saying what he/she said" is often not a straightforward, but a modified or a qualified occurrence of a speech act verb, or even a full and detailed paraphrase of his/her utterance. Perhaps we should grant that the very possibility of performing a certain action lies, not strictly in the linguistic means for saying that it is or was performed (describing, reporting it, and the like), but (at least as much) in the linguistic and paralinguistic means for showing (suggesting, indicating, and the like) that it is or was performed. On the whole, to admit of a correspondence between socially available (linguistic) actions and lexically available (linguistic) action verbs could push us to extremely structural-functional conclusions, to the effect that every individual can in each occasion do only one among a finite number of socially pre-determined things (or speech acts). To recognize a role for showing, suggesting, indicating, and for the implicit understanding they bring about could help us to account for the fact that the cultural storehouses of action descriptions are not fixed once for ever but are, like everything that is human, bound to evolve.

3. Speech Acts As Actions

There is no point in talking of speech acts unless the perspective suggested by such a terminology has a heuristic value of its own. Otherwise, we may as well limit ourselves to, say, a traditional functions-of-language terminology. Now, according to one accepted meaning of 'act', 'speech act' can be taken to mean a piece of speech activity. This is rather uninteresting, since nobody would contend that speaking is an activity, even a social one, at least in the weak sense of an 'active' (and not 'passive') condition of human individuals. To use the term 'speech act' in this sense is superfluous. But we also claimed that acts can be considered as actions. If this can be said of speech acts too, a less trivial perspective is open to us.

However, to consider speech acts within the framework of a causal theory of action is again not enough. If they are actions merely in the sense that they are caused by (linguistic) intentions and other pro-attitudes, speech act theory turns out to be identical with a theory of speaker's meaning. But the latter could well dispense with any speech act terminology.

The view proposed in 2.2. is better suited to satisfy the requirement that speech act theory should have an object of its own. According to this view, to consider speech acts as actions amounts to claiming that an agent/speaker who produces an utterance may be viewed as bringing about by his/her utterance a state St_2, by way of contrast to a presupposed state St_2. The states that speech acts bring about are to be found (broadly speaking) in the context. Therefore, to grant that speech acts are actions is to allow for context change. And as soon as speech acts are envisaged as action, bringing about changes in their own contexts, speech act terminology ceases to be superfluous. It is given the task of drawing attention to how speech affects its context, as well as to a number of related questions, such as how the context-changing dimension of speech is to be connected with 'meaning', 'language structure', and other main topics in linguistic or logical theory.

3.1. Is the speech act one action or a collection of actions? Since its very beginning, speech act theory has been concerned with subdivisions of the speech act, such as the locutionary, the illocutionary, and the perlocutionary act. Nevertheless, Austin insisted that "The total speech act in the total speech situation is the *only actual* phenomenon which, in the last resort, we are engaged in elucidating" (Austin (1975, p. 148)).

Such a question forces us to face the problem of action identity. Very sketchily, there are two ways out. We could adopt a low-level identity criterion (like the one adopted by Davidson (1980, p.59)) and identify actions by means of the corresponding 'basic' actions (physical movements). All actions performed by performing the same 'basic' action will be considered as descriptions of the same action. Or we could adopt an opposite high-level criterion (like the one adopted by Shwayder (1965, p. 37, pp. 47-48) with respect to typal identification). Actions will be identified by identifying their purpose, or better, in our view, their result. The relevance of a certain purpose (or, in our view, result) to our present interests will determine which action we are focusing attention upon. Other acts performed together with the main action will be considered as part of the description of the way in which the action is performed. If, with reference to them, the language of action is used, they are to be called 'collateral' acts (cf. Shwayder (1965, p. 53, pp. 133 ff)). This view has also something in common with the "fine-grained" proposal for act individuation by Goldman (1970), except that in Goldman's view it is not possible to consider collateral acts as included in one main action.

A low-level identity criterion is uninteresting for speech act theory, since basic physical movements are not what the speech act theorist is interested in. Such movements cannot by themselves define what kind of act the agent/speaker is performing. Think for example of Austin's example "Iced ink" (a case of homophony; cf. Austin (1975, p. 124)), or of the illocutionary ambiguities of almost any isolated utterance.

The high-level criterion is more akin to our interests. In the interactional view adopted here, it would be convenient to think of action identity as relative to the result St_2 that it is relevant for P_2 to put under P_1's responsibility. Action identity is in this case far less stable than in the case of the low-level criterion, since in real life the relevant results are often quite complex ones, and relevance itself can shift. While the use of an action verb often presupposes certain collateral acts and/or includes within the action a chain of successfully achieved consequences, all these things can in turn be talked about in an act-identificatory language, so that the identity of the main action seems to waver (is it identical with one of its collateral acts? or to a collection of them?). I suspect that such a context-relative unstableness, though unpleasant for logicians, is something the speech act theorist must resign him/herself to. We should be prompt to consider the total speech act as an action and its collateral acts as the way in which the action is performed; but in addition, whenever it becomes relevant to shift attention onto one of the collateral acts as being an action itself, we should apply our considerations about actions to it. In doing so, we should not worry about counting actions once and for all: the enumeration of the St_2s which are the case in a given situation, if made according to one judgement about what is relevant, is simply not comparable to an enumeration made according to another such judgement.

3.2. One might still wonder why the "total speech act in the total speech situation" is never tackled directly in speech act theoretical considerations.

Searle's identification of the illocutionary act with the (whole) speech act (cf. Searle (1969, pp. 45 ff.)) seems to provide a comfortable answer to this question. It allows us to take into account the illocutionary act together with its propositional content, and to think that such a consideration is exhaustive with respect to the speech act. But while the description of an illocutionary act may involve the description of the (collateral) locutionary act, it is not always so. In most cases, many features of the locutionary act are not taken into consideration. Moreover, we ought to allow for non-linguistic equivalents of illocutionary acts. And in general, should we really grant that the total effect of a speech act is no more and no less than the effect of the illocutionary act? It would be preferable to view both the locutionary and the illocutionary acts as collateral to the "total speech act" that puzzles us.

The main trouble here is that the "total speech act" does not seem to allow for descriptions other than the descriptions of its own collateral acts. And a description consisting in a set of partial descriptions can never be exhaustive. Even a full and well-concerted paraphrase loses at least certain acoustic and rhythmic effects.

Austin outflanked this difficulty by focusing attention not on the positive features of speech acts, but on the ways in which they fail. If something is faulty, it is so in one respect or in another and not in a total, all-embracing way. Therefore, a negative approach provides insights into particular features of the action which may possibly count as subordinate, collateral acts. If we accept that failures of collateral acts are also partial failures of the action, these insights can throw some light also on the latter.

Of course, the issue is not settled. All I am suggesting here is that no locutionary, illocutionary, or perlocutionary act tells us the whole story about the "total speech act", and that such acts should not be added up, but should be considered each against the background provided by the full speech act as is presented to us in perception and (implicit) understanding.

3.3. We have by now established that it is possible to mark out subdivisions in the total speech act. But is it worth-while to do so, especially with respect to the locutionary-illocutionary-perlocutionary distinction? Such distinctions are relevant to a discussion of speech acts as actions only in so far as they are accompanied by parallel distinctions among kinds of context change, so that each subdivision is characterized by the bringing about of a particular kind of context change.

Ballmer (1983) distinguishes "physical", "mental", "social", and "linguistic" context changes. His list is full and detailed, since each group is subdivided into smaller categories, but it is not free from ambiguities and overlapping. For example, some social context changes involve mental ones, while some mental context changes could well count as social ones. I do not want to propose an alternative grouping. Instead, I would like to put forward a pre-requisite for any possible grouping of context changes, by suggesting a methodological distinction between a 'material' and a 'semiotic' level in the manifold notion of 'context'. This distinction does not take into account (as do distinctions such as the one between 'co-text' and 'context') the form of the *expression* of the so-called contextual elements, that is, whether these become manifest to the participants through linguistic representation or by being given to perception and cognition. What it takes into account is the form of the *content* of contextual elements, and in particular whether they are meant to have a *de facto* existence (either of a physical or of a psychological kind), or the somewhat *de iure* existence of semiotic constructions. The ambiguities and overlapping present in Ballmer's list of context changes are often due to mixing up the 'material' and the 'semiotic' context.

The proposed distinction can be successfully applied to the locutionary-illocutionary-perlocutionary subdivision. Illocutionary acts appear to bring about strictly 'semiotic' context changes. Perlocutionary acts appear to bring about various 'material' context changes. Locutionary acts appear to be linked both to some kinds of 'material' context changes (including physical ones), and to some kinds of 'semiotic' context changes (expression and/or construction of meaning, introduction of referents, and the like).

A consequence of all this is that we should give up any all-embracing speech act classification. There are no positive features of the speech act as a whole that could be taken as a classificatory criterion. If on the contrary we go back to the speech act's subdivisions, the features of the locutionary, the illocutionary, and the perlocutionary act can only provide classificatory criteria for the field of context changes that is covered by each subdivision. Therefore, any classification of, say, illocutionary acts requires the previous delimitation of the field of illocutionary context changes.

4. Illocutionary Context Change

According to Austin (1975, pp. 116-18), illocutionary acts are linked to effects in three different ways.

(1) They must "secure uptake": that is, the utterance must be understood by a hearer as having a certain meaning and a certain force.

(2) They "take effect" in certain ways, which are to be distinguished from "producing consequences in the sense of bringing about states of affairs in the 'normal' way, i.e. changes in the natural course of events".

(3) They "invite a response": for example, an order invites the response of obedience, and a promise that of fulfilment.

In our search for a specifically illocutionary kind of context change, the kind of effects in (2) is the most obvious candidate.

4.1. Effects in (2) above have sometimes been characterized as 'conventional'. But the use of such a word needs some comments, if not justification. First of all, we ought to establish where the supposed 'conventionality' lies.

Austin (1975) claims that both performative utterances and illocutionary acts are 'conventional'. Such a claim, perhaps because of the widespread interest in explicit performative formulas, has generally been understood as referring to the *means* for performing the act rather than to the achieved *effect*. In Searle (1969), conventional felicity conditions are convertible into semantical rules for the use of illocutionary force indicating devices. Strawson (1964) redefines the conventionality of illocutionary acts in terms of Gricean non-natural meaning, thus favouring the identification of speech act theory with the theory of speaker's meaning. Such approaches lead to denying the illocutionary act any conventionality, apart from the well-known conventionality of language, and are radicalized in more recent approaches which give an inferential analysis of illocutionary force (Leech 1983) and even of locutionary meaning (Bach and Harnish 1979). Certainly, authors such as Bach and Harnish admit of a restricted number of "conventional" illocutionary acts, that belong to institutional contexts and are performed by satisfying a convention; but, again, such acts are deemed to be conventional basically because of the conventionality of the means by which they are performed.

If, on the contrary, we are concerned with the conventionality of an action, and if the notion of an action is analyzed along the lines suggested above, what is relevant is to see whether the state St_2 brought about by the action is itself conventional. Now, that this is so is suggested, in Austin (1975), by the fact that both the means and the effects of the act are established by the existence of an "accepted conventional procedure" (cf. felicity condition A.1., Austin (1975, p. 14, p. 26)); and, in Searle (1969), by the fact that the essential condition, which defines the point of the act, is included among the conventional conditions for the successful performance of illocutionary acts.

Moreover, the degree of defeasibility of illocutionary acts is higher than that of many non-linguistic actions and even of some linguistic ones. Not only can the speaker's responsibility for the act be extenuated or cancelled, but the very state St_2 can be cancelled and replaced by St_1, or by some state St_2', St_2'', St_2'''... for which the agent/speaker will be held responsible instead of for St_2. Think, for example, of what may happen when a subordinate addresses a rough imperative sentence to a superior. Only conventional states can be cancelled, as opposed to their material context consequences: think of the children born from a 'void' marriage, or of the thousand times she ironed her alleged husbands' shirts.

Now, it could be maintained that 'conventionality' is at work in illocutionary acts, but only in so far as they act upon institutional states. All non-institutional cases ought to be considered non-conventional ones. Here, I would like to introduce an element that is perhaps fundamental in all 'conventions' (inclusive of 'institutions'): agreement (and/or negotiation). Of course in most cases of 'convention' the participants' agreement is not explicitly declared nor discussed, but is achieved in implicit ways, e.g. through complex mutual expectations (like in Lewis (1969)). If, as Austin claims, the hearer's uptake is necessary for the illocutionary act to count as successfully performed, and if such an uptake includes the understanding of the force of the utterance, a St_2 can be successfully brought about by the illocutionary act only in the case where the participants *agree* about which St_2 it has to be. Without such an agreement no stable 'conventional' effect takes place.

Of course there may be lots of different cases. If the addressor and the addressee disagree openly, and no further socially accepted criterion is at hand in order to decide who is right, no conventional St_2 is brought about. If the disagreement is in-apparent and a conventional St_2 (but not the same one) is held by both to be the case, the alleged illocutionary act is likely to be questioned and revised, or even retracted, in subsequent interaction. It is in such ways that the interactional relationships set up by performing illocutionary acts are 'voidable'.

The conventional nature of the illocutionary act has also been recognized by such authors as O. Ducrot, D. Wunderlich, G. Gazdar. Ducrot (1972) argues that the illocutionary act has "juridical" effects on the speaker's and the hearer's rights and obligations. According to Wunderlich (1976), the illocutionary act introduces or retracts interaction conditions, among which norms, social values and the like are also included. Wunderlich also suggests that such changes in interaction conditions should be represented by deontic logic. Gazdar (1981) proposes to define speech acts as functions from contexts into contexts and incorporates into context not only alethic or epistemic, but also deontic propositions. It should be noted, moreover, that even reference to alethic or epistemic propositions is mediated by the commitment of the speaker, which is a deontic notion, as in the example: "an assertion that Φ is a function that changes a context in which the speaker is not committed to justified true belief in Φ into a context in which he is so committed" (Gazdar (1981, p. 69)).

Here, I would like to attempt the description of illocutionary context changes in informal, and fairly ordinary deontic language. I shall list some prototypical examples, inclusive of recognizedly non-institutional cases.

An order changes the context so that the addressee is assigned a new obligation. A question changes the context so that the addressee is assigned the obligation to answer, i.e. to give the addressor some relevant information. A permission changes the context so that an obligation or prohibition is retracted from the addressee, and the addressor is assigned the obligation to refrain from interventions on the addressee's behaviour within the scope of the permission. A warning changes the context so that from the addressee the right is retracted to hold the addressor responsible for something that may affect him/her in the situation referred to by the warning. A promise changes the context so that the addressor is assigned a new obligation, and the addressee is assigned the right to expect a certain subsequent behaviour on the part of the addressor. A statement changes the context so that the addressee is assigned the right to state in his/her own turn what the addressor stated and the addressor is assigned the obligation to give, if requested, some evidence for the statement. An apology changes the context so that from the addressor an obligation acquired by violating the addressee's rights is retracted, while from the addressee the right to initiate aggressive behaviour towards the addressor is retracted.

It may be seen very easily (apart from the further refinements needed) that all the above-mentioned changes are 'semiotic' context changes. They cannot be identified with 'material', *de facto* states of affairs, but with *de jure* semiotic constructions. Moreover, to borrow an useful term from narrative semiotics, they can be represented as changes in the "modal competence" of the participants: that is, in the set of modal (here restricted to deontic) predicates, that belong to each participant at each stage of the ongoing interaction (for an investigation of this perspective, cf. Sbisà (1984)).

4.2. The first kind of effect mentioned in 4., i.e. the bringing about of the hearer's uptake, should not be identified with the result of the illocutionary act. It does not coincide with that St_2, which is the core of our analysis of illocutionary acts as actions. The understanding of the utterance as having a certain meaning and a certain force is, rather, an effect of the locutionary act, brought about by means of the lexical, syntactical, morphological, and intonational properties of the utterance. It involves both changes in the material context (the speaker's voice reaches the hearer's ears; a processing of the perceived sounds takes place, involving cognitive operations, emotional reactions, etc.) and changes in the semiotic context (the 'knowledge state' of the hearer: there are some propositions that the hearer is now entitled to know).

However, the 'uptake' seems to be a two-fold phenomenon. One side of it consists in the connection between the speaker/hearer relationship and the hearer's understanding of the speaker's utterance. The hearer may trust, or distrust the speaker; be confident or suspicious; assume that the cooperation principle (cf. Grice (1975)) is at work, or that the situation is a 'transformed' one (whether 'keyed' or 'fabricated': cf. Goffman (1974)). A positive, confident expectation state favours the understanding of the speaker's force showing devices at their literal value, provided that these do not clash with the assumption that the cooperation principle is at work. A negative, suspicious attitude may favour indirect and even totally opposite interpretations.

Such attitudes need, of course, further analysis. The role they play is undoubtedly puzzling: they can be represented within the semiotic context (the cooperation principle works as a premise for conversational inferences; Goffman's keyings and fabrications are transformations of frame; Greimas' *contrat fiduciaire* is just a 'contract'), but at the same time they are anchored in the bio-psychological history of the hearer, which is a part of the material context. Therefore, even if the dynamics of uptake could be completely clarified, no actual uptake would ever be predictable on the sole basis of a semiotic description. Conversely, each actual uptake can be considered as a 'given' starting point for building up a whole semiotic-context machinery.

The other aspect of the uptake phenomenon is its relationship to illocutionary context changes. We may be tempted to say that the uptake is a belief on the part of the hearer to the effect that a certain illocutionary act has been performed. But if the uptake plays, with respect to the illocutionary act, the crucial role we described in 4.1., so that the hearer has 'the last word' about the speaker's performance, such a claim is too weak. No *belief* can make itself true. But, after all, we say that someone *knows* something on those very occasions in which we accept the presupposition that he/she draws on 'the truth' about that matter; and in this case we admit precisely that the hearer is entitled to establish the truth about the speaker's performance. Therefore, I prefer to say that, unless there are reasons to cast doubt on a particular hearer, the hearer knows (and not merely believes) what the speaker has brought about. Of course such a knowledge (like any type of knowledge) can be extended or modified, and even questioned or proved false; nevertheless, unless this happens, it is (real) knowledge, though partial and relative to a point of view. It might also be claimed that 'knowledge' is a social and deontic-like concept (perhaps even a defeasible one), while 'belief' embodies a stronger claim about the believer's psychological reality. But here the material context is not so relevant as it was with respect to the other aspect of the uptake phenomenon we discussed previously.

4.3. As far as the fact that illocutionary acts 'invite a response' is concerned, we should not be persuaded by this to give up the illocutionary-perlocutionary distinction. Responses are effects of the illocutionary act in the sense of the *Erfolgreichsein* discussed by Wunderlich (1976, pp. 115-18). They stem from the felicity of the illocutionary act (which we consider as one kind of successfulness specific to illocutionary context change) in connection with some other features of the situation. As far as the response is considered as 'invited' by the illocutionary act, it is an effect or consequence of it, though not in a strictly causal sense (since it also depends on a decision on the part of the hearer). It remains, however, at the same time a further action (whether linguistic or not) on the part of the hearer: the latter may now be represented as a P_1, to be held responsible for bringing about a St_2 by contrast to a presupposed St_1. When precisely this St_2 is considered as being brought about by the speaker, who performed the speech act that has invited the response, or under his/her responsibility, we say that a perlocutionary act takes place.

Between the illocutionary act and the response to it, as well as (since the response can be described as the St_2 of a perlocutionary act on the part of the speaker) between the illocutionary and the perlocutionary act, there is a gap which is due once again, as in the case of the uptake, to the anchoring of the speech episode in the bio-psychological history of the participants. Of course, which is the 'right' response to each type of illocutionary act is a matter of convention, or in other words, it follows from the illocutionary act's own definition. But whether an actual order achieves obedience, an actual promise is fulfilled or an actual apology succeeds in eliminating a grudge, is not a matter of agreed interpretations or of assigning modal predicates. Something must occur in the material context, outside, though not independently of the illocutionary act. The relevant factor here could be described as the 'alignment' or 'contraposition' of the hearer to the perlocutionary goals of the speaker (these concepts are akin to the notion of "acceptation" proposed by Wunderlich (1976, pp. 116-17) as the second step of the *Erfolgreichsein*). Of course, in order to tackle the issue of alignment in a more detailed way further reflection is needed, especially about the expressive dimension of illocutionary (and possible of locutionary) acts.

4.4. By way of a summary, I propose a tentative schema of the dynamics of the illocutionary act.

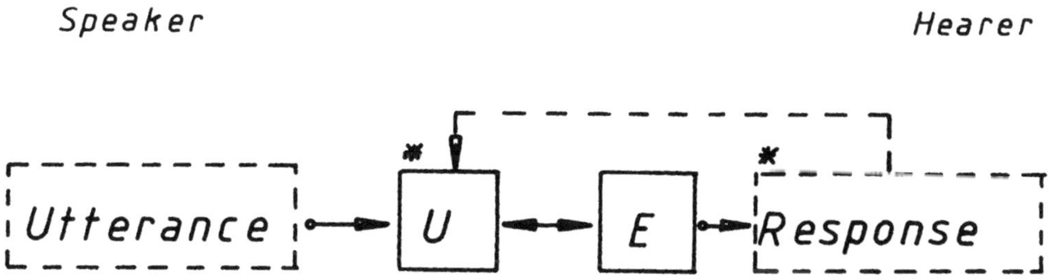

Fig. 2

Note that:

(1) The utterance produces an illocutionary context change E only if it produces an uptake U.

(2) The achieving of E, i.e. a (conventional) St_2 considered as being brought about by the speaker in contrast to a presupposed (conventional) St_1, is one and the same thing as the successful performance of an illocutionary act.

(3) The double arrow between U and E is meant to indicate that as soon as U is achieved (in the simplest case, as soon as the hearer identifies which E has to take place and accepts the speaker's performance as felicitous), E is achieved as well; and that, conversely, any achieved E presupposes being duly recognized in U.

(4) The dotted lines and arrows indicate what is open to direct observation: an utterance by a speaker and a response by a hearer. It must be remembered that the hearer's response is often the only means available to other participants for deciding whether and how the hearer understood the speaker's illocutionary act. Other participants can infer from the hearer's response, together with some information (or assumption) about the speaker/hearer relationship at the two levels of trust-or-distrust and of alignment-or-contraposition, how the illocutionary act was understood and what illocutionary context change was brought about. This topic is, however, well known to conversational analysts.

(5) The asterisks indicate the two crucial places, already discussed in 4.2. and in 4.3., where unpredictability comes in.

5. Context Change and Pragmatics

Up to this point, the reader may still wonder whether by introducing context change in pragmatics we are not simply introducing another terminology. By way of conclusion, I would like to argue that there is much more at stake than a different way of talking.

What we are proposing is to reconsider the relationship between context and language. In Austin's time, to bring context into the theory of language was still a rather audacious task. After all the work done in pragmatics in the past twenty years, both by linguists, logicians and philosophers of language, claims such as these: all use of language is a use in a context; the meaning and the force of utterances are disambiguated only in the context; not everything can be said in every context not for merely 'grammatical' reasons but for 'pragmatic' ones; have by now become shared background assumptions. But this is not enough. A pragmatics for which language is merely determined by context would be unable to account for any signification exceeding what is already given (let alone linguistic *action*). It would also be an inescapably synchronic pragmatics, avoiding any problem related to the possible evolution of linguistic and pragmatic rules.

Sociological discussions have shown clearly enough that context is no *deus ex machina* and that it cannot be considered as distinct from the ways in which the participants deal with it. It is now time to explore the other aspect of the relationship between context and utterance: that is, to investigate the ways in which language defines, enlarges, and modifies its context. The considerations in this paper are aimed at suggesting the guidelines for such investigations. Our natural ally, which for practical reasons has been ignored in this paper, is the textual approach to semiotics which sets up context (inclusive of addressor and addressee) through the consideration of textual features (cf. for example Greimas, Courtês (1979, pp. 125-28)).

Of course a great deal of topics of undoubted relevance to the construction of a dynamic pragmatics have not even been mentioned here. I am thinking, for example, of 'presuppositions' and other kinds of implicit signification. I did not even clarify the role of the 'presupposed' St_1 in my account of illocutionary acts. It seems to me that there is a wide and complex class of effects connected to speech acts, which do not involve a change in the context subsequent to the speech act, but a change in the accepted description of the context previous to the speech act. Such backward effects, among which many 'presuppositions' should be included, deserve special attention both for their peculiar 'conventional' character and for the help that their investigation could give to our understanding of such phenomena as micro-changes in linguistic and cultural rules.

By considering the dynamic side of the language/context relationship, pragmatics would cease to resemble a post-modern linguistic *Galateo*, and might perhaps succeed in developing into a theory of human interaction, not without practical interest itself, inasmuch as it could be an instrument for interaction both at the individual and at the social level.

Bibliography

Austin, J.L., 1970, Philosophical Papers, Second Edition, ed. by J.O. Urmson and G.J. Warnock, Oxford University Press, London.

Austin, J.L., 1975, *How to Do Things with Words*, Second Edition, ed. by J.O. Urmson and M. Sbisà, Oxford University Press, London.

Bach, K. and R.M. Harnish, 1979, *Linguistic Communication and Speech Acts*, M.I.T. Press, Cambridge, Mass. and London.

Ballmer, Th.T., 1982, 'Context Change and Its Consequences for a Theory of Natural Language', in *H. Parret* et al. (eds.), *Possibilities and Limitations of Pragmatics*, John Benjamins, Amsterdam, 17-55.

Ballmer, Th.T. and W. Brennenstuhl, 1981, *Speech Act Classification*, Springer Verlag, Berlin.

Beardsley, M.C., 1978, 'Intending', in A.I. Goldman and J. Kim (eds.), *Values and Morals*, Reidel, Dordrecht, 163-84.

Brennenstuhl, W., 1975, *Handlungstheorie und Handlungslogik*, Scriptor Verlag, Kronberg.

Comrie, B., 1976, *Aspect*, Cambridge University Press, London.

Davidson, D., 1980, *Essays on Actions and Events*, Oxford University Press, London.

Ducrot, O., 1972, Dire et ne pas Dire, Hermann, Paris.

Feinberg, J., 1964, 'Action and Responsibility', in M. Black (ed.), *Philosophy in America*, Allen and Unwin, London, 134-60.

Garfinkel, H., 1967, *Studies in Ethnomethodology*, Prentice-Hall, Englewood Cliffs, N.J.

Gazdar, G., 1981, 'Speech act assignment', in A.K. Joshi, B.L. Webber and I.A. Sag (eds.), *Elements of discourse understanding*, Cambridge University Press, Cambridge and London, 64-83.

Goffman, E., 1974, *Frame Analysis*, Harper and Row, New York.

Goldman, A.I., 1970, *A Theory of Human Action*, Princeton University Press, Princeton, N.J.

Greimas, A.J., 1966, *Sémantique structurale*, Larousse, Paris.

Greimas, A.J., 1970, *Du Sens*, Seuil, Paris.

Greimas, A.J., 1983, *Du Sens II*, Seuil, Paris.

Greimas, A.J. and J. Courtés, 1979, *Sémiotique: Dictionnaire Raisonné de la Théorie du Langage,* Hachette, Paris.

Grice, H.P., 1975, 'Logic and Conversation', in P. Cole and J.L. Morgan (eds.), *Syntax and Semantics: Speech Acts,* Academic Press, New York and London, 41-58.

Hart, H.L.A., 1951, 'The Ascription of Responsibility and Rights', in A. Flew (ed.), *Logic and Language*, I Series, Blackwell, Oxford, 145-66.

Leech, G.N., 1983, *Principles of Pragmatics,* Longman, London.

Lewis, D.K., 1969, *Convention*, Harvard University Press, Cambridge, Mass.

Parret, H., M. Sbisà and J. Verschueren (eds.), 1982, *Possibilities and Limitations of Pragmatics,* John Benjamins, Amsterdam.

Petöfi, J.S., 1974, *Semantics, Pragmatics, Text Theory* (Working Papers, Series A, n. 36), Centro Internazionale di Semiotica e Linguistica, Urbino.

Sbisà, M., 1984, 'On illocutionary types', *Journal of Pragmatics* 8, 93-112.

Sbisà, M. and P. Fabbri, 1980, 'Models (?) for a Pragmatic Analysis', *Journal of Pragmatics* 4, 301-19.

Searle, J.R., 1969, *Speech Acts,* Cambridge University Press, London.

Searle, J.R., 1975, 'A Taxonomy of Illocutionary Acts', in K. Gunderson (ed.), *Language, Mind and Knowledge* (Minnesota Studies in the Philosophy of Science, 7), University of Minnesota Press, Minneapolis, MN, 344-69.

Shwayder, D.S., 1965, *The Stratification of Behaviour,* Routledge and Kegan Paul, London.

Strawson, P.F., 1964, 'Intention and Convention in Speech Acts', *The Philosophical Review* 73, 439-60.

Von Wright, G.H., 1968, *An Essay in Deontic Logic and the General Theory of Action,* North Holland, Amsterdam.

Ware, R., 1973, 'Acts and Actions', *Journal of Philosophy* 70, 403-18.

Wunderlich, D., 1976, *Studien zur Sprechakttheorie,* Suhrkamp, Frankfurt a.M.

PART THREE: LANGUAGE PROCESSES WITHIN DIFFERENT FRAMEWORKS (LOGICS; ARTIFICIAL INTELLIGENCE; STATISTICS)

THE DYNAMIC INTERPRETATION OF DISCOURSE

Alice G.B. ter Meulen
Rijksuniversiteit Groningen

0. Preface

The body of this paper was written in 1981-82, when the use of partial models for the semantics of natural language was only beginning to gain some popularity, due to innovative research by Hans Kamp, Jon Barwise and John Perry and Irene Heim. In the subsequent years mathematical methods for semantics have been developed which supersede the formalization of semantic interpretation outlined in this paper. The paper thus can be read as marking a historical phase in the semantics of natural language: the transition from Montague Grammar with its complete information models to dynamic interpretation based on partial information models.

1. Introduction

In the current linguistic research on discourse phenomena, various research disciplines seek a new, common territory, a new set of puzzles and problems and a shared perspective on these. This paper is intended as a contribution to these developments from modeltheoretic semantics for natural language, research rooting in the tradition of Montague Grammar. The characterization with modeltheoretic tools, derived from mathematical logic, of the notions an *interpretation of a discourse, logical* or *pragmatic consequences* from a discourse, and *coherence* of a discourse is the ultimate goal, taking intuitive judgements about truth, soundness of inferences in natural language and incompatibility of information as empirical starting-point. The main theoretical question directive of this research is which model-structures out of all logically possible ones are best suited to the interpretation of natural language, representing formally how we convey information using natural language.

In contrast to Montague's adagium that there is *no* important difference between natural and formal languages,[1] which licenced his universalizing strategy in semantics, the characteristic linguistic concern now is to single out just those semantic properties that characterize all and only the human, natural languages. Along these lines, linguistically significant universals about semantic properties of all natural languages can be formulated in terms of abstract properties of mathematical models.[2]

The use of partial models outlined in this paper has been guided to some extent by the important ideas of Situations Semantics, a modeltheoretic semantics recently developed jointly by Jon Barwise and John Perry.[3] But the fragment is here drastically impoverished by excluding sentence-embedding verbs, indexical expressions and tense and aspect. The simple, extensional, partial models are intended to demonstrate 'in vitro' what is to be gained from using partial rather than total functions in models that are gradually built up while processing the information contained in the sentences of the discourse. The grammar gives a *direct* interpretation of discourse in the language of the fragment by assigning to the linguistic expressions set-theoretic constructions and hence avoiding the translation to an intermediate formal language. Models are defined by a set of instructions for the construction of semantic objects, contrary to the commonly used truth-definitions. In some respects, the partial models in this paper are

---

1 See the opening statement of Montague's 'English as a Formal Language' in Montague (1974), for instance. Typical of this strategy was the method of raising everything to an intensional type, and using a convention plus a meaning-postulate to get the desired extensional inferences.

2 Barwise and Cooper (1981) provide such universals and offer a general perspective of quantification, which proves to be extremely useful for these linguistic questions. See also Van Benthem (1983).

3 See Barwise (1981), and Barwise and Perry (1981a),(1981b),(1983).

close to possible world semantics, in particular to Kripke frames with a partial order on the reference-points and expanding domains. Although solving the intricate problems of belief, intensionality and indexicality may require us to abandon possible worlds and the accompanying set-theoretic reconstruction of intensional notions of properties and propositions, the present paper has the limited ambition to show that by just releasing the assumption of complete information a modeltheoretic account can be given of the referential and attributive use of definite descriptions, originally introduced into the philosophical debate on reference by Keith Donnellan in various papers.[4] Partial models can be taken to represent states of knowledge, or models of interpreted information, representing what a speaker takes to be true about the world. The differences in the information various people may have can be represented as differences in the strength of their information. The referential use of a description is stronger than the attributive use in constraining the possible continuations of the interpretation in ensuing discourse. The notion of strength of information is characterized in modeltheoretic terms below.

In modeltheory the need for formal representation of information about a structure was recognized at least twenty years ago, and, since P. Cohen's results on the independence of the Continuum Hypothesis, the powerful technique of forcing is widely used to construct models of arbitrary inductive theories. The stepwise construction of models for consistent sets of sentences may prove a fruitful heuristic for the semantics of natural language and the dynamics of interpretation. However, in the semantics for the fragment of this paper sentences are instruments for constructing models, but once they have performed that function, their interpretation does not necessarily remain constant or stable in expansions of the constructed model. This feature will be discussed and illustrated later. The relation between partial models and forcing techniques for finite theories is not further discussed

---

4 See Donnellan (1966) and (1968).

here.[5]

After the preliminary discussion of the linguistic motivation and the main notions of the partial models, a fragment is presented in which the interpretation of a discourse in the language of the fragment is defined. It is indicated why partial models are more useful than models based on complete information. The fourth section presents the modeltheoretic characterization of referential and attributive uses of descriptions, concluding with a brief discussion of some other proposals made in the literature on descriptions. The final section contains a brief outline of pronoun interpretation in partial models, employing witness-sets, and a discussion of unbound anaphora cases.

2. Partial Models

It has often been argued that a necessary, but not sufficient condition for interpreting a sentence of a natural language is knowledge in what conditions the sentence is true. Making a statement by uttering a sentence is an invitation to the audience to construct a representation of the world in which these conditions are satisfied and to consider it to be a representation of what is actually the case (neglecting for the moment all complicating factors like tense and attitudes). The instrumental character of expressions of natural language has been emphasized before in Game Theoretic Semantics in terms of winning strategies in interpretive games (Saarinen (ed.), 1979), and context change models of Ballmer (1981), for instance. People use what they hear or see to change what they know about the world and to understand how other users of the language see reality. Different people may well construct different interpretations of one and the same sentence, depending on what they already believe is the case and on their prejudices or predispositions. But for all these inter-

---

5 For an expository and encyclopedic account of forcing see Barwise (1977).

pretations to be models of what is said, they must share many important features. A sentence contains information depending on what context it is used in and who interprets it. In isolation of context and speaker a sentence is interpreted by the set of partial models in which it is true. A discourse, taken as a set of sentences ordered by the natural order of utterance-time, is interpreted by a family of such sets of partial models, that meets certain requirements. When the discourse is continued and more information is added, this family of sets of models is accordingly restricted, while the partial models are developed further, and become richer as representations of the given information. In this sense, information plays a double role, one of restricting the set containing all possible models for the sentences of the discourse, and secondly, stating the conditions on expansions of any given partial model in the set of possible models.

Adding new information may require three different kinds of elaboration of a given partial model: i) introduction of new individuals into the domain, ii) storing new information about already available individuals, or iii) revising the model to accommodate the new information. The second kind of information is usually conveyed with dependent pronouns. The fragment in this paper contains only singular personal pronouns, interpreted as referring to individuals. The account in this paper of the dependencies between NPs and singular anaphora is only a rudimentary sketch of the main dynamic procedures of interpretation.

3. A simple fragment

In the fragment determiners are assigned a lexical category, which makes the interpretation of NPs compositionally dependent upon the head noun, in accordance with the interpretation of NPs as generalized quantifiers. NPs that are constructed from proper names and pronouns are syntactically marked by an asterisk, reminicent of a similar practice in Montague Grammar, to distinguish them syntactically from their homophonic basic expressions. The usual connectives are entered in the lexicon, except for negation which is treated syncategorematically for reasons that are explained later. The discussion of the fragment is continued after its presentation.

SYNTAX

LEXICON

Proper names	(PN)	-	*Janet, Jim*
Pronouns	(PRO)	-	*she, he*
Nouns	(N)	-	*murderer, woman, man*
Intrans, verbs	(IV)	-	*walks*
Transitive verbs	(TV)	-	*is, kisses, loves*
Determiners	(D)	-	*the, every, some, a*
Connectives	(C)	-	*and, or, if ... then*

Let T, the set of *basic terms*, abbreviate PN U PRO and let P, the set of *property-expressions*, be N U IV.

SYNTACTIC RULES

S1. If $\alpha \in$ D and $\beta \in$ N then $\alpha\beta \in$ NP.

S2. If $\alpha \in$ T then $\alpha^* \in$ NP.

S3. (i) If $\alpha \in$ IV then *does not* $\alpha' \in$ IV, where α' is the infinitive of α, and α is a basic lexical item.

(ii) If $\alpha \in$ TV and $\alpha =$ *is* then *is not* $\in$ TV and if $\alpha \neq$ *is* then *does not* $\alpha' \in$ TV, where α and α' are as above.

S4. If $\alpha \in$ NP and $\beta \in$ TV then $\beta \in$ P.

S5. If $\alpha \in$ NP and $\beta \in$ P then $\alpha\beta \in$ S.

S6. If $\alpha, \beta \in$ S then α *and* $\beta \in$ S.

S7. If $\alpha\ \beta \in$ S then α *or* $\beta \in$ S.

S8. If $\alpha\ \beta \in$ S then *if* α *then* $\beta \in$ S.

Let S be the set of well-formed sentences defined recursively in this syntax. A *discourse* in the fragment is any finite set of sentences $D \subseteq S$, linearly ordered by the natural, temporal order of utterance, i.e. $D = \langle S_0, S_1, S_2, \ldots, S_n \rangle$, where every $\langle S_o, \ldots S_{m-1} \rangle$ is the *preceding discourse* of S_m.

SEMANTICS

In the semantics for this fragment we define the *interpretation of a discourse* inductively, by constructing partial models for the sentences of the discourse unless a contradiction in the discourse is arrived at. In that case the construction of a partial model for the discourse is preceded by a revision of the given model, discarding the least information that makes the model incompatible with the new information. Since we contradict ourselves frequently without being incomprehensible, interpreting what is said requires flexibility and a cooperative strategy, i.e. dropping some of the previously acquired information, in order to accommodate what is now taken to be correct information. Contradictions serve a specific function in the dynamics of interpretation, as we will see below.

The following set of rules define the construction rules for a model in the interpretation of the discourse. From an external point of view the models are partial in two respects: i) they may contain only part of the entire domain of individuals, and ii) they may contain only partial information (positive or negative) about the introduced individuals. From the syntactic analysis of each sentence in the discourse, a model for the discourse is obtained by a stepwise construction based on the order of the sentences in the discourse. Let S_0 be the first simple, i.e. non-composite, sentence. We construct a model for it following these instructions. Let I be a supply of discourse-referents, from which we choose referents for the interpretation of NPs; $[[\,\alpha\,]]$ indicates the interpretation of α. Let a model $\underline{M}$ consist of a subset of I, called I_M and $[[\;]]_{\underline{M}}$, i.e. $\underline{M} \langle I_{\underline{M}}, [[\;]]_{\underline{M}} \rangle$.

(i) assign every $\alpha \in T$ in S_0 one discourse-referent, selected from the witness-sets available at $\underline{M}$ if α is a pronoun:[6] $[[\alpha]]_{\underline{M}} \in I_{\underline{M}}$.

(ii) assign every $\alpha \in P$ some subset of $I_{\underline{M}}$: $[[\alpha]]_{\underline{M}} \subseteq I_{\underline{M}}$.

(iii) assign every $\alpha \in TV$ some subset of pairs of $I_{\underline{M}}$: $[[\alpha]]_{\underline{M}} \subseteq I_{\underline{M}} \times I_{\underline{M}}$.

The other expressions in S_0 contribute to the construction of a model according to the following rules.

SEMANTIC RULES

M1. (i) $[[\textit{every}\ N]]_{\underline{M}}$ construct $\{X \subseteq I_{\underline{M}} \mid [[N]]_{\underline{M}} \subseteq X\}$

(ii) $[[\textit{some}\ N]]_{\underline{M}}$ construct $\{X \subseteq I_{\underline{M}} \mid \overline{[[N]]_{\underline{M}} \cap X} = 1\}$[7]

(iii) $[[\textit{the}\ N]]_{\underline{M}}$ construct $\{X \subseteq I_{\underline{M}} \mid [[N]]_{\underline{M}} \subseteq X \text{ and } \overline{[[N]]_{\underline{M}}} = 1\}$

(iv) $[[\textit{a (n)}\ N]]_{\underline{M}}$ construct $\{X \subseteq I_{\underline{M}} \mid \overline{[[N]]_{\underline{M}} \cap X} = 1\}$

M2. $[[T\ *]]$ construct $\{X \subseteq I_{\underline{M}} \mid [[T]]_{\underline{M}} \in X\}$

M3. (i) Assign $[[\textit{does not}\ P]]_{\underline{M}} \subseteq I_{\underline{M}}$ such that $[[P]]_{\underline{M}} \cap [[\textit{does not}\ P]]_{\underline{M}} = \emptyset$

(ii) Assign $[[\textit{does not}\ TV]]_{\underline{M}} \subseteq I_{\underline{M}} \times I_{\underline{M}}$ such that $[[TV]]_{\underline{M}} \cap [[\textit{does not}\ TV]]_{\underline{M}} = \emptyset$, and idem for $[[\textit{is not}]]_{\underline{M}}$

M4. $[[\beta\,\alpha]]_{\underline{M}}$ construct $\{x \in I_{\underline{M}} \mid \{y \mid \langle x,y\rangle \in [[\beta]]_{\underline{M}}\} \in [[\alpha]]_{\underline{M}}\}$

M5. $[[\alpha\,\beta]]$ construct $[[\beta]]_{\underline{M}} \in [[\alpha]]_{\underline{M}}$

6 Pronoun interpretation using witnesses is explained in the last section of this paper.

7 $\bar{\alpha}$ indicates the cardinality of α .

These rules for construction of models allow as interpretation of a set of sentences many different models, disjoint in domain and different in background information obtained from previously interpreted sentences. The *interpretation* of S_0 is the set of models that can be constructed by means of these rules.

To continue the interpretation of the discourse, the next simple sentence S_1 is used to construct an elaboration or expansion of the model constructed for S_0. Using the same instructions the set of models for S_1 is obtained. The composition of the two basic sentences is obtained according to the rules M6-8. But two preliminary notions need to be defined first.

A partial model $\underline{M}$ can be expanded to a 'larger' partial model $\underline{M}'$ by adding new individuals and assigning wider extensions to the basic lexical items.

Definition $\underline{M} = \langle I_{\underline{M}}, [[\ ]] \rangle$ *is expanded* to $\underline{M}' = \langle I_{\underline{M}'}, [[\ ]]_{\underline{M}'} \rangle$ iff. $I_{\underline{M}} \subseteq I_{\underline{M}'}$ and $[[\ ]]_{M} \subseteq [[\ ]]_{\underline{M}'}$ for all basic lexical N, IV and TV.

Accordingly, $\underline{M}$ is called a *submodel* of $\underline{M}'$ when $\underline{M}$ is expanded to $\underline{M}'$.

A simple sentence α is *true* on the basis of $\underline{M}$ when $[[\ ]]_{\underline{M}}$ is defined and constructed in M. Once a sentence is used to construct a model, it does not necessarily remain true in every expansion of this model.

To the three syntactic construction rules for complex sentences correspond these three semantic rules.

M6. Construct $[[\alpha \textit{ and } \beta]]_{\underline{M}}$ by constructing $[[\alpha]]_{\underline{M}}$ and expand $\underline{M}$ to $\underline{M}'$ such that $[[\beta]]_{\underline{M}'}$.

M7. Construct $[[\alpha \textit{ or } \beta]]_{\underline{M}}$ by expanding $\underline{M}$ to some $\underline{M}'$ such that $[[\alpha]]_{\underline{M}'}$ or $[[\beta]]_{\underline{M}'}$.

M8. Construct $[[\textit{if } \alpha \textit{ then } \beta]]_{\underline{M}}$ by constructing in every $\underline{M}'$expanding $\underline{M}$ in which $[[\alpha]]_{\underline{M}'}$ also $[[\beta]]_{\underline{M}'}$:

In this way the models constructed for basic sentences are used compositionally to construct models for complex sentences.

The construction of the set of models for sentences of the fragment is now defined recursively, but how do we use the rule in continuing the interpretation of the discourse? A discourse as defined earlier consists of a sequence of sentences. A partial model constructed for the first sentence of the discourse is used and expanded to a model for the next sentence, and sentence boundaries are interpreted as sentential conjunction. Since every positive noun or verb is assigned an extension disjoint from its negative counterpart, it is impossible to construct a model for a discourse containing a contradiction. When the instructions conflict with what is already assigned, revise the model by reducing it to its maximal submodel that is compatible with the new information. The process of revision may not be deterministic, since there may at any one stage be more than one 'maximal' submodel compatible with the new information. Various pragmatic factors as well as other semantic constraints enter into the choice of the 'best revision', when contradictory information is given, but it is outside the scope of this paper to elaborate on such issues.

The reason for treating negation syncategorematically in this fragment is that, normally, in exchanging information about the world we state that we believe is the case as well as what we take not to be the case, keeping track of positive and negative extensions of property phrases in this way. By analyzing negation syncategorematically, a compositional analysis is given up, but that, by itself, is not a strong argument against it. In specifying semantics for a language we are in principle free to chose what expressions to treat as logical constants and what expressions as denoting expressions. Our choice depends upon what we wish to account for in the interpretation of the language. Instead of working with positive and negative extensions of property-phrases we could have introduced a notion of falsification, using artifact Falsum,[8] but this would certainly not be any more intuitive or realistic.

Note that in a partial model it is not required that every element in its domain is listed either in the positive or in the negative extension of every property-phrase interpreted in that model. Elements of the domain are in this sense only partially described, until the final stage is reached in which they are included in either the extension of every property-phrase of the language or in its negative counterpart. A model is *complete* if it assigns an individual to every name of the language and lists each on either positive or negative extension of every property-phrase of the language. Such a model is reminiscent of a Carnapian state-description and comes close to an isolated possible world. At times reference is made to such a complete model with $\underline{M}^+$, and we assume that $\underline{M}^*$ is the complete true model of the world.

Conjunction is interpreted by constructing the model for the left conjunct, which is expanded to a model for the second conjunct. Conjunction in natural language often indicates such a dependency between the constituent sentences, making conjunction an asymmetric relation between sentences.[9]

Disjunctive information commits us to the condition that in every path leading from $\underline{M}$ to $\underline{M}^+$, i.e. in every way we may obtain new information, at some stage $\underline{M}'$ we are bound to reach the information that one or the other of the disjuncts is true. Disjunctive information commits us by conditions on expansions of the model we are using for our interpretation.

Conditional sentences are interpreted in a strong way here,

---

8 See Veltman (1981) which uses an 'impossible fact' and falsification for a propositional language.

9 This asymmetry is often argued for with sentences like

He put his pyjamas on and fell asleep
*He fell asleep and put his pyjamas on.

requiring that for every expansion in which the interpretation of the antecedent is constructed the consequent must be constructed as well. As a result the consequent of a conditional represents information that is at most as strong as the information represented in the antecedent. The fragment could be extended with a biconditional connective, if desired, which, accordingly, were to be interpreted by requiring that both sentences are interpreted in exactly the same set of models. Expressions that are in this strong sense equivalent represent equivalent information. A rule of substitution could subsequently be formulated allowing substitution of such equivalent expressions in every context with preservation of its informative content. Although this may be a useful addendum, it will not be explored here any further.

Alternative proposals on the interpretation of conditionals abound in the recent literature.[10] The definition of strong consequence in Barwise and Perry (1983) is similar, although it is formulated in Situation Semantics, which takes situation-types, or event-types as basic notions. Here partial models are outlined only in their barest skeleton to show what is gained from having partial functions construct interpretions of sentences rather than total ones. The interpretation of conditionals is one such gain, it represents more adequately the way conditionals are used than the standard material implication.

The fragment contains only a few determiners, interpreted as constructing generalized quantifiers over the domain. These determiners were chosen to illustrate some of the semantic properties of generalized quantifiers, and to characterize the notions of definiteness and indefiniteness of NPs. The determiner *the* is interpreted as the principal filter generated by the singleton,

---

10 The state of the art is well represented in the special issue on conditionals of the *Journal of Philosophical Logic* (1981), vol. 10, no. 2, and in *Ifs*, W.L. Harper et.al. (eds), Reidel Publ. Co., Dordrecht, 1981, and in E. Traugott et al (eds), (1986), *On Conditionals*, Cambridge University Press.

which is the denotation of the noun in the description, as in Barwise and Cooper (1981). If the extension of the noun already contains more than one individual, the interpretation of the determiner *the* and hence of the entire definite description is not defined. This is in accordance with the intuition that we do not understand who is being referred to by *the man*, when we have already been talking about two men before. In such a case we require more specific information describing which man is meant, so that the individual referred to can be singled out uniquely. When the denotation of the noun is subsequently extended in an expansion of the model in which the description was interpreted, we cannot continue to refer to the same individual with a description containing only the same noun, since uniqueness will then be violated. The sentence containing the description served its purpose of providing the instructions for reference to a unique individual, but it is no longer interpreted in its expansions. If we wish to continue to refer to the same individual we must use a proper name or an anaphoric pronoun bound by the description as antecedent. In that case we use the description referentially. Identifying the referent of the description by a proper name or pronoun or demonstrative is characteristic of a referential use of the description. In attributive use we are not able to identify the referent of the description in a direct connection to the model we are constructing. We choose an individual, a discoursereferent as the referent of the description, but do not identify him any further. I.e. no sentence of the form

The N is *a*

is interpreted in the same partial model, where $a \in T$. This is the basis of the account of referential and attributive descriptions, discussed in more detail in the next section.

The two determiners *a* and *some* are in their semantical behavior virtually identical, as we excluded *some* with plural nouns. They are indefinite determiners that show characteristic behavior in expansions of partial models.[11]

The universal determiner *every* is restricted to the finite domain of individuals of the model in which it is interpreted. From a logical point of view it is undesirable to be restricted to finite domains, but using universal quantifiers in natural language we rarely employ a countable domain, as our outlook is inherently finite. Sentences containing universal NPs are not preserved in every expansion of a model in which they are interpreted, since by adding new individuals they may range only over a subset of the domain of the expansion. A simple dialogue can illustrate this point.

A: 'Every woman kisses Jim.'
B: 'But Janet does not kiss Jim'.
A: 'Well, then not every woman kisses Jim. Janet is a woman and she does not kiss him.'

The model for this dialogue first constructs a set of women, and then the generalized quantifier $\underline{Q}$, the family of sets in $T_{\underline{M}}$ that contain the set of women. Next the set of individuals kissing Jim is constructed such that it is a member of $\underline{Q}$. B introduces a new individual Janet, constructing $I_{\underline{M}}$, which according to our conventions concerning proper names and gender, is understood to be included in the set of women, extending the basic assingments. Then the referent for Janet is included in the set of individuals that do not kiss Jim, disjoint from the set of individuals that do. Now A, quantifying over $I_{\underline{M}}$, can consistently deny her earlier assertion. The anaphoric pronouns are interpreted by the referents that are available in the appropriate witness-sets. The selection of such referents is discussed in the final section of this paper.

This dialogue shows that we often quantify over restricted domains and revise our statements as soon as we are confronted with a

---

11 They are persistent, i.e. for all A, $B \subseteq I$ if $A \subseteq B$, then $D(A) \subseteq D(B)$. See Barwise and Cooper (1981) and Van Benthem (1981).

counterexample to our claim by extending the domain. But if such a simple, inductive view of universal quantification is not considered appropriate for natural language interpretation, we may allow functions into the domain of individuals, or obtain the effect of universal quantification over potentially infinite domains by putting conditions on the expansions of the basic nouns and verbs that occur in the quantifiers. It may well be that we use universal quantification both in the finite and in the infinite interpretation, which would have the two interpretations coexist in an interpretation.

4. Interpreting descriptions

Descriptions can be interpreted in two essentially different ways. Referentially used descriptions refer directly to an individual, and in expanding models we keep this reference constant. Attributively used descriptions require a unique referent, but we have no means of direct reference to single him out. The semantic connection between a referent of a referentially used description and the linguistic expression is just like the connection between a name and its referent or a pronoun with a proper name as antecedent. Knowing the murderer is knowing e.g. that Jim is the murderer, but knowing that the murderer is a man, we do not know who the murderer is or how to fix the referent of the description in a stable way.

It is sometimes remarked in the discussion on descriptions that the referential use provides maximal information about the referent, whereas attributive use provides only weaker information. But it is difficult to characterize exactly how much information is sufficient to establish the direct link between an individual and a description for it to be an instance of referential use. In some cases we seem to know very little about a directly accessible referent, e.g. 'The stranger next to me wears a red coat'. In other cases we still use a description attributively even though we know much about the referent, as in the final stages of a search for the murderer. Therefore, attempting to account for the difference in terms of *amount* of information about the referent is not a feasible approach. But if we take *strength* of information to characterize the difference we can say that a referential use provides stronger information:

Definition: In a model $\underline{M}$ an expression *provides stronger information than an expression* , when for more expansions $\underline{M}'$ contain $[[\alpha]]_{\underline{M}'}$ than $[[\beta]]_{\underline{M}'}$. Information is stronger when more expansions preserve the interpretation, i.e. when the interpretation is less sensitive to the addition of information. This allows us to characterize the strength of the information provided by a referentially used description. Every model in which a description is interpreted referentially also interprets a sentence of the form

The P is *a*

where a $\in$ T, a proper name or pronoun. In other words, a referential description strongly implies a sentence in which the referent is fixed by a directly referential term. We commonly understand descriptions in such a referential way, not only committing the user to know that a referent exists, but also to be able to refer directly to him using proper name or pronoun (bound or deictic). Only in cases in which there is a common assumption that the referent is not yet fixed in the information available, or in cases in which such specific information is irrelevant to the purpose of the communication, descriptions are used attributively. Often attributive descriptions contain functional information, i.e. information that specifies only a function by which the referent can be found when the function is supplied with an appropriate argument - when more information is available. For instance, a description 'The winner of this race', when used before the race is started, can only be used attributively. Of course, we intend to refer to the same individual with the attributive description, all the time speaking about him before the race. In this respect attributive descriptions refer rigidly, to the same discourse-referent, although his identity is not known yet, and no direct reference can be made to him.

A model in which a description is interpreted referentially also provides an interpretation for 'the P is *a*', even though this sentence did not necessarily occur in the discourse. The interpretation of the description is fixed so that it is insensitive to the addition of further information. Extending the interpretation of P loses the conditions for the interpretation of such sentences, but then the name *a* is already introduced and remains fixed up till $\underline{M}^*$ is reached. Attributively interpreted descriptions do not yield models in which such a reference-fixing sentence is interpreted as well. Until direct reference to the referent is provided, we gain more information about him by interpreting sentences of the form

the P is Q

where Q contains only quantified NPs and verbs. By gaining more information we get a fuller description of the referent, until in $\underline{M}^*$ the attributive use provides a complete description, identifying the referent with the set of all of its properties.

Donnellan pointed out in his papers that although we may use a description referentially, we may find out later that the property used in the description is not actually true of its referent. This claim has provoked much bewilderment in the discussion on reference and descriptions. We understand the situation to be as follows: when we construct a model for a sentence with a referential description, we also have to interpret a sentence in the same model that identifies the referent and we intend to keep the interpretation of that sentence stable in expanding the model subsequently. When we are confronted with a sentence contradicting that stable sentence, we have to retrace our construction of the model till we reach the submodel in which that sentence was introduced as new information. From that model we drop the information provided by that sentence and add the new stable sentence and consistently expand that new model to contain all the information we originally had, when we encountered the contradictory information. In other words, we trace back on our path through the partial order induced by possible model extensions, until we reach the point where the identifying stable sentence was added, and rebuild the model with the new stable sentence, following another path in that partial order. In general, revision of our model amounts to tracing back to the largest submodel that is consistent with our new information and rebuildung a model containing all the information available with the new sentence added.

This account of the interpretation of descriptions does not require two distinct syntactic analysis-trees for the two uses of descriptions. There is only one syntactic analysis of a sentence containing a description in the fragment, and the referentially used description is interpreted by a subset of the set of models by which the attributive description is interpreted. The process of interpretation is, however, entirely compositional. There is no reason to require that sentences that can be interpreted differently depending on the context in which they are used should have two or more distinct syntactic analysis trees. Such a homomorphism between syntax and semantics, characteristic of Montague Grammar, is too strong a theoretical restriction for a semantic based on direct dynamic interpretation.

The account of the ambiguity in the use of descriptions is similar to what was proposed informally in Johnson-Laird & Garnham (1980). They suggested that the two uses should be distinguished by the knowledge which a speaker intends to be relevant to the interpretation of his utterance. But they allow a wider criterion for referential use including the cases in which a speaker knows the referent of the description *the P* is identical to the unique referent of *the Q*, for at least one distinct predicate Q. It seems, however, that this condition is too liberal. We may well use two distinct descriptions both in an attributive way, and know that their referent is one and the same individual, without knowing who he actually is. In the following setting this case is clearly demonstrated: 'The winner of the race is the one who runs fastest'. Direct reference to the referent by means of a pronoun or proper name seems to be the correct condition on models in which a description is interpreted referentially.

In Klein (1981) referential descriptions are discussed in the context of *de re* belief reperts. The condition that referential descriptions impose in the interpretation in the framework of this paper is

$$[[x = \textit{the}\ \zeta]]\ {}^{M}_{ca}[u/x]$$

where *a* assigns individuals to variables independently of the world coordinate, and in a directly referential way, and *c* is the context. This proposal captures part of our criterion, although it is couched in a possible worlds model. But it leaves out that we may well use a description referentially when we interpret in the same model a sentence with a proper name identifying the referent of the description. Proper names and referential pronouns function in this respect in the same way. They both fix the referent in a way which makes it insensitive to the addition of information.

Finally, there is the account for the referential and attributive use of descriptions in situation semantics, see Barwise & Perry (1983). The difference is expressed in the elements of the situation-types that form the interpretation of the sentence with the description. In referential use, called the value-loaded use, the individual itself is a constituent of the situations that make the description true. In the attributive use the property used in the description is the constituent of the situations that interpret the description in its value-free use. Both individuals and intensional entities like properties can equally be constituents of situations, from which situation-types can be abstracted. But this semantic realism regarding properties is not required to account for just the difference in use of descriptions, when partial models are allowed and strength of information can be defined on model-expansions.

5. Interpreting pronouns

If there is a single most important problem in discourse-semantics it must be the interpretation of pronouns. The fragment presented here contains only singular pronouns, which simplifies the situation considerably. Still, there are too many sides to the interpretation of pronouns to account for in this paper. In this section some indication is given of the main line of thought in the account of pronoun interpretation by referential assignment of appropriate referents of the domain. It should allow for a uniform treatment of deictic pronouns, bound pronouns and so called unbound anaphora that occur in the wellknown 'donkey-sentences'.

Pronouns in this fragment are interpreted on a par with proper names as maximal filters insensitive to the addition of further information. They are stable in every model-expansion. A model constructed for previous discourse is to determine which referent is to be selected as the interpretation of the pronoun as basic lexical item. A simple example illustrates the main idea:

Jim walks. He is a man.

First a model $\underline{M}$ is constructed for the first sentence, in which the family of subsets of $I_{\underline{M}}$ that contain Jim, has the set of walkers as member. The next sentence produces an expansion $\underline{M}'$ of $\underline{M}$ in which *he* is to be interpreted as coreferential with *Jim*. The two filters that are the interpretation of these two expressions are generated by identifying their generator as the same singleton containing Jim. The extension of the property of being a man is then constructed as a set which is a member of the filter interpreting the pronoun.

Agreement of gender is, of course, a first requirement for co-reference. This could be imposed on the pronoun interpretation rule in the fragment, by simply listing gender of pronouns and marking every name for gender.

In Barwise and Cooper (1981) the notion of a *witness-set* for a generalized quantifier is defined as follows:

Definition A *witness-set* for a quantifier D(A) living on A is any subset w of A such that $w \in D(A)$.

The singleton Jim is thus a witness-set for $[[\, Jim \,]]_M$. For indefinite NPs a witness-set is any non-empty set in the interpretation of the property-expression of the NP. Since we required singularity in the construction of the interpretation of indefinite NPs, such witness-sets contain exactly one member with the relevant property. Definite NPs have ony one witness-set, indefinite NPs may have as many as the cardinality of the head-noun extension in the model is (disregarding negative NPs).

In interpreting a pronoun we may take any witness-set of the generalized quantifier that is its antecedent, and the element of the witness-set is the referent of the pronoun. This is a general procedure for finding the right referent in a partial model that is assigned to the pronoun in that model.

When more than one referent is available for pronoun interpretation, there are corresponding ambiguities in the sentences themselves and various possible antecedents for the pronoun.

Notice that a sentence like
Every man walks. He kisses Janet.

is correctly generated by the fragment, but is not interpretable with an anaphoric reading of *he* or *every man*. The unique witness-set for the definite *every man* is the set containing all men, but *he* requires a single referent for its interpretation, hence it is not interpretable in such cases. The conditions on coreference are stated as part of the rule of pronoun interpretation and the notion of a witness set. If no appropriate witness set can be found on basis of the interpreted previous discourse, the pronoun is to be interpreted deictically. Deictic gestures then constitute another means of introducing an individual into the domain of the model.

An example of pronoun interpretation with an indefinite NP is the following.

A woman walks. She kisses Jim.

The indefinite NP *a woman* walks is interpreted by the class of subsets of the domain that contain exactly one referent. Consequently, a witness-set based on this quantifier contains exactly one woman, which serves as the referent of *she* in the next sentence, when the model is expanded to be a model of the two sentences.

Both definite NP, which include proper names as well as demonstrative determiners (not in this fragment) and definite descriptions, and, on the other hand, indefinite NPs yield witness-sets that contain the referents for the interpretation of pronouns. There is no need to distinguish the relation between a definite NP and a pronoun from the relation between an indefinite (or otherwise quantified) NP, as argued for so often in current theories on binding and anaphora.

A major problem for pronoun interpretation was formed by so called 'donkey-sentences' and unbound anaphora.[12] In the present fragment similar sentences are generated, consider this sentence.

If Janet kisses a man, she loves him.

According to the interpretation of *if ... then* sentences in this fragment, whenever in a model-expansion $\underline{M}'$ of the present model $\underline{M}$ the antecedent *if*-clause is interpreted, the consequent must be interpreted too. Take in any such $\underline{M}'$ a witness set of $[\![\text{Janet}]\!]_{\underline{M}'}$, and the element of that set is the referent of the pronoun *she*. In every $\underline{M}'$ *a man* is interpreted by the class of subsets of the domain that contain exactly one man, and from this we can construct a witness set which determines the single

---

12 Unbound anaphora are not c-commanded by their antecedents, crossing sentence boundaries. See Hintikka and Carlson (1979) and Kamp (1981).

referent, an individual of the domain I' assigned to the pronoun *him*. By employing the notion of a witness-set based on the generalized quantifier that is the antecedent for the pronoun, the account of pronoun interpretation is quite straightforward. The universal force of the indefinite description is due to the conditional, and should not be imposed on the general interpretation of the indefinite NPs in all contexts. Indefinite NPs in conditionals are interpreted by requiring that in any expansion that interprets the antecedent, i.e. in which some individual exists that interprets the noun of the NP, the consequent, in which anaphoric reference is made to that individual, is interpreted as well. The universal force is not located in the NP, but in the interaction of the NP and the conditional. A sentence like

If Janet kisses every man, she loves him.

cannot be interpreted with a coreferential reading of *every man* and *him*, since the witness-set for a universal NP contains at least two elements and hence does not provide a single referent to be the interpretation of the pronoun *him*. The sentence itself is, of course, generated in the fragment, and it can be interpreted when a witness-set is found based on an NP interpreted in a submodel, that does yield a single referent. Otherwise the pronoun must be interpreted deictically.

Note that definite NPs in conditionals are interpreted in a way not fundamentally different from indefinite NPs.

If Janet kisses Jim, she loves him.

is interpreted with the unique referent for *him* determined by the witness-set for the generalized quantifier, the ultrafilter generated by the proper name *Jim*. Hence it is not necessary to give an independent account of pronoun interpretation in 'donkey'-sentences for indefinite and definite NPs. The same basic strategy applies in every pronoun interpretation.

Although the discussion of pronoun interpretation in this section is a far cry from giving a complete and formally precise account of all the issues in this complex topic of semantic interpretation, the main features of such an account have been outlined. The notion of a witness-set provides the crucial tool for this account based on generalized quantifiers and partial models and expansion thereof. It demonstrates that a more procedural approach in model-theoretic semantics is a fruitful strategy for anophora and problems of coreferentiality.

4. Conclusions

In this paper a simplified fragment was given with a semantics for discourses in this fragment. It was illustrated how models are constructed for sentences and expanded to models for sequences of sentences. The gain of working with partial models was demonstrated in the account of the referential and attributive account of definite descriptions, an issue commonly thought of as pragmatic. The interpretation of pronouns based on generalized quantifiers generated by witnesses from appropriate witness-sets is only outlined, but along these lines a more comprehensive account of pronoun interpretation will give a truly semantic solution to many problems of anaphora that resist syntactic explanation.

Bibliography

Ballmer, Th., 1981, 'Context change and its consequences for a theory of natural language', In Parret et.al. (eds), *Possibilities and limitations of Pragmatics*, John Benjamins, Amsterdam (1981).

Barwise, J. (ed.), 1977, *Handbook of Mathematical Logic*, North-Holland Publ. Co., Amsterdam

Barwise, J. 1981, 'Scenes and other situations', *Journal of Philosophy* LXXVIII, 7, July (1981), p. 369-397.

Barwise, J. and R. Cooper (1981), Generalized Quantifiers and Natural Language', *Linguistics and Philosophy* 4, (1981), p. 159-219.

Barwise, J. and J. Perry, 1981a, 'Semantic innocence and uncompromising situations', in French et.al. (eds), *Midwest studies in Philosophy VI*, University of Minnesota Press, Minneapolis.

Barwise, J. and J. Perry, 1981b, 'Situations and attitudes', *Journal of Philosophy LXXVIII*, November (1981), p. 668-691.

Barwise, J. and J. Perry (1983), *Situations and attitudes*, MIT Press, Cambridge.

Van Benthem, J., 1983, 'Determiners and Logic', *Linguistics and Philosophy* 6: 4, p. 447-479.

Donnellan, K., 1966, 'Reference and definite descriptions', *The philosophical review* 75 (1966), p. 281-304.

Donnellan, K., 1968, 'Putting Humpty-Dumpty back together again', *The philosophical review* 77 (1968), p. 203-215.

Hintikka, J. and L. Carlson, 1979, 'Conditionals, generic quantifiers, and other applications of subgames', in Margalit (ed.), *Meaning and Use*, Reidel Publ. Co, Dordrecht, (1979), p. 57-92.

Johnson-Laird, P.N. and H. Garnham, 1980, 'Descriptions and Discourse Models', *Linguistics and Philosophy* vol. 3, no. 3, (1980), p. 371-393.

Kamp, H., 1981, 'A theory of truth and semantic representation', in Groenendijk et.al. (eds), *Formal Methods in the Study of Language*, Part 1, Mathematical Center Tracts 135, Amsterdam, (1981), p. 277-322. Reprinted in *Truth, information and interpretation*, same eds, GRASS 2, Foris Publications (1984).

Klein, E., 1981, 'Defensible descriptions', in Heny (ed.), *Ambiguities in Intensional Contexts,* Reidel Publ. Co., Dordrecht (1981), p. 83-102.

Montague, R., 1974, *Formal Philosophy,* Collected papers edited by R. Thomason, Yale University Press, New Haven (1974).

Partee, B., 1972, 'Opacity, coreference and pronouns', in Davidson and Harman (eds), *Semantics of Natural Language,* Reidel Publ. Co., Dordrecht (1972).

Saarinen, E. (ed.), 1979, *Game Theoretic Semantics,* D. Reidel Publ. Co., Dordrecht.

Veltman, F., 1981, 'Data Semantics', in Groenendijk et.al. (eds), *Formal Methods in the Study of Language*, Part 2, Mathematical Center Tracts 136, Amsterdam, 1981, p. 541-564. Reprinted in *Truth, information and interpretation,* same eds, GRASS 2, Foris Publications (1984).

USER MODELLING IN ANAPHORA GENERATION: ELLIPSIS AND DEFINITE DESCRIPTION

Anthony Jameson and Wolfgang Wahlster
University of Hamburg

Abstract

This paper shows how user modelling can improve the anaphoric utterances generated by a dialogue system. Two kinds of anaphora are examined: contextual ellipsis and the anaphoric use of singular definite noun phrases. In connection with ellipsis generation, anticipation of the way in which the user would be likely to reconstruct a given utterance can help to ensure that the system's utterances are not so brief as to be ambiguous or misleading. When generating noun phrases to characterize specific objects with which the user is not familiar, the system may take into account the existential assumptions, domain-related desires, and referential beliefs ascribed to the partner. These applications of user modelling are illustrated as realized in the dialogue system HAM-ANS, and some possible generalizations and extensions of the strategies described are discussed.

Research on HAM-ANS is currently being supported by the German Ministry for Research and Technology (BMFT) under contract 081T1503 8.

Introduction

Many writers on dialogue processes have pointed out ways in which it can be advantageous for a participant in a dialogue to be aware of the beliefs, desires, and goals of the other participant and to anticipate the partner's response to one's own utterances.

The importance of the user modelling principle has been demonstrated with its implementation in several AI systems (e.g. Cohen 1978, Allen 1979, Rich 1979; cf. also Jameson/Hoeppner/Wahlster 1980); the present paper focuses on possible roles for user modelling in the generation of utterances which must be interpreted against the background of previous utterances: The types of anaphora to be considered are contextual ellipsis and definite noun phrases.

Anticipation Feedback in the Generation of Elliptical Utterances

Many of the natural language AI systems developed to date have included the capability to process elliptical utterances of the user's (among the more extended discussions of this topic are those of Grosz 1978 and Kwasny/Sondheimer 1981). As far as the generation of elliptical utterances is concerned, the state of research presents a paradox: On the one hand, most systems whose output takes the form of natural language utterances at all (as opposed, e.g., to lists, tables, or graphic displays), generate utterances which must be regarded as elliptical. But on the other hand, the generation of elliptical utterances has seldom (e.g. by Simmons/Slocum 1972) been viewed as a problem in its own right; elliptical utterances either follow predefined templates or verbalize data structures representing attributes, object names, or other nonpropositional information which have been returned by the part of the system responsible for determining the content of the system's utterance.

When the problem of ellipsis generation is treated explicitly, one of the questions which arises is that of how elliptical an utterance can be without becoming ambiguous or misleading. Consider for example, a dialogue system which plays the role of a hotel manager answering on the telephone the questions of a prospective guest about a particular room:

(1) User: Are there three beds and at least one desk in the room?

If the room in fact has three beds and three desks and the system is to mention the latter fact explicitly instead of simply answering 'Yes', it might respond

(2) System: Yes, three.

But the user is likely either to interpret 'three' as referring to the number of beds or to be uncertain as to the correct interpretation.

The slightly less elliptical utterance
(3) System: Yes, three desks.

is still somewhat confusing, as it suggests that the system may be answering the question with respect to the desks only while avoiding a comment about the number of beds.

Complete clarity is provided only by an answer which includes considerable overlap with the question:

(4) System: Yes, three beds and three desks.

But how is the system to know that (2) and (3) would be too elliptical? The following example shows that no general rules relating the form of an elliptical utterance to that of the relevant previous utterance are likely to provide sufficient guidance in all cases.

If the room has three beds and two desks, and the system answers the question (1) with

(5) System: Yes, two.

no uncertainty of the sort caused by (2) can arise, since the possibility that 'two' here refers to the number of beds is ruled out (given the fact that the answer is prefaced by the word 'yes').

It is difficult to conceive how a system could take into account the factors which make (5) acceptable but (2) unacceptable except by anticipating how the user is likely to interpret its utterances.

This conclusion should come as no surprise, since analogous processes have frequently been postulated in connection with other aspects of language production, such as the avoidance of referential and other ambiguities, the control of nuances, and the correction of mistakes in the speaker's own utterances after they have been articulated (see, e.g., Hoenkamp 1980 and Winograd 1977).

The term *anticipation feedback loop* will be used in the following to refer to the type of control algorithm which involves the simulation of the user's response to an utterance of the system's. In order to provide a concrete example of how such a control algorithm can be realized in a natural language dialogue system, we will now illustrate the operation of the ellipsis generation component of the system HAM-ANS, which, in one of its roles, simulates the hotel manager of the examples on page 311 (cf. Jameson et al. 1980). The generalizability of some of the features of this component will then be discussed.

Ellipsis Generation in HAM-ANS

The structures which represent the content of utterances to be generated by HAM-ANS are always complete formulas of the logic-

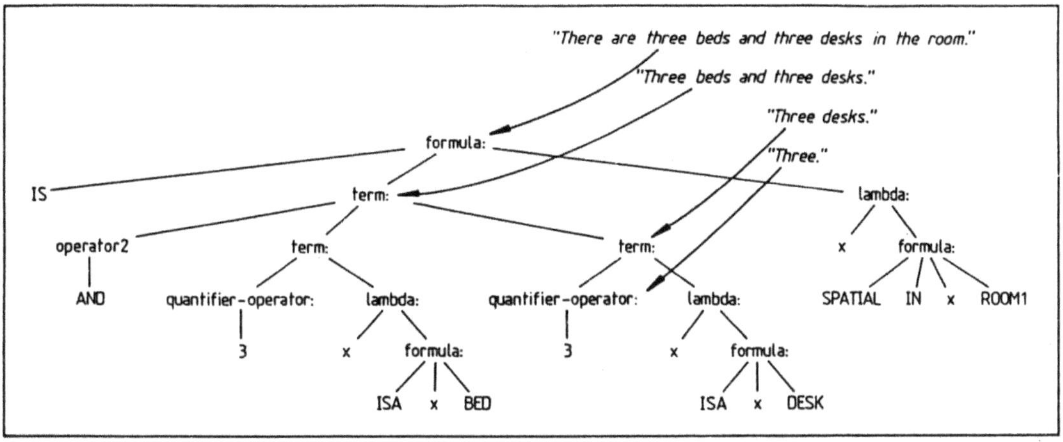

Fig. 1: Structure of a formula representing an answer to the question 'Are there three beds and at least one desk in the room?'

based representation language used which express propositions. Because they often overlap considerably with the previous utterance, such formulas are not in general suitable for complete verbalization. On the other hand, there are two advantages that result from basing all language generation on structures representing propositions: (a) Such structures can be processed by the system in various other ways in addition to being verbalized. For instance, they can be added to the part of the user model which consists of the facts assumed to be known to the user. (b) The task of selecting appropriate elliptical formulations can be assigned to a specialized ellipsis generation component which is in a position to deal with complications of the sort illustrated by the examples presented above.

The first step in generating an utterance is the transformation of the formula to be verbalized so that its structure corresponds much more closely to the syntactic structure of the corresponding complete sentence. For example, the answer to (1) is transformed into the formula which is sketched in Figure 1 as a tree structure. A formula which has been transformed in this way has the property that each of its subtrees corresponds to a syntactic constituent of the corresponding natural language utterance. Since the ellipsis generation component is restricted to the generation of elliptical utterances which correspond to single, complete syntactic constituents of the given formula (other types of ellipsis will be mentioned in the next section), the subtrees of the transformed formula thus form the set of structures from which possible elliptical reductions of the formula can be selected.

The actual selection of one such subtree (which may be the entire formula, if no elliptical reduction is possible) is performed by the ellipsis generation component. The formula to be reduced is compared to a stored representation of the previous utterance - in the present example a formula corresponding to (1) which is essentially identical to the structure of Figure 1 except for the subtree representing 'three' in the term for 'three desks'.

Analysing the formulas for both the answer and the question from the top downward, the system makes a list of the successively smaller subtrees (pointed to by the arrows on the right in Figure 1) of the answer formula which can be obtained by eliminating immediate subtrees which are essentially identical in the two formulas. The smallest of the possible reductions found (here: 'three') represents the most concise possible verbalization of the answer which doesn't leave out any information not already contained in the question; but there is no guarantee that it will be interpreted correctly by the user. Rather than verbalizing it immediately, the system therefore attempts to reconstruct it as the user would, i.e. by determining how it fits into the structure of the original question.

In simulating the user's reconstruction attempt, the system activates the same ellipsis reconstruction component that it uses when it itself is interpreting an elliptical utterance of the user's. (This component is in many ways similar to the one described by Grosz 1980). It looks for subtrees within the representation of the previous utterance which are similar in structure and content to the elliptical structure. If there is one such subtree which is clearly a better candidate than all other subtrees, the elliptical structure is substituted for this subtree in the representation of the previous utterance in order to obtain a formula which presumably represents the meaning of the ellipsis. If there are either no such subtrees or two or more approximately equally good candidates, the ellipsis is considered not to be (uniquely) interpretable.

Applying this reconstruction strategy to the possible ellipsis 'three', the system finds as possible corresponding parts of the question not only the subtree representing 'at least one', but also the one representing 'three' within the term for 'three beds'. Since neither of these is judged to be clearly preferable to the other according to the similarity metric used, it cannot be assumed that the user would interpret the answer 'three' correctly.

The ellipsis reconstruction strategy is therefore applied to the next smallest of the possible elliptical structures, the one representing 'three desks'. Again two possible corresponding parts are found: not only the intended subtree, the term for 'at least one desk', but also the conjoined term for 'three beds and at least one desk'. The simpler subtree is judged to be a somewhat more plausible candidate than the conjoined term, being more similar in structure to the hypothetical elliptical utterance, but the difference in plausibility is not great enough to meet the system's standard. It therefore calls the ellipsis reconstruction component a third time, applying it to 'three desks and three chairs'. This matches the term for 'three desks and at least one chair' in the question formula so closely that it is judged to be unambiguously interpretable and thus suitable for verbalization.

The Generalizability of This Approach

It is important to note how simple the anticipation feedback loop described in the preceding section is: When a candidate utterance is found to be ambiguous, no diagnosis is made of the ambiguity as a basis for the selection or construction of the next candidate; the next candidate is simply taken from the list of possibilities supplied by the ellipsis generation procedures. This kind of simplicity is only possible in a local anticipation feedback loop, i.e. one in which the generation procedures for one particular aspect of an utterance are linked with the recognition procedures for the same aspect. An example of a more global loop, in which error diagnosis and the generation of alternatives are necessarily more complicated, is the control structure sketched by Hoenkamp (1980).

HAM-ANS's generation and recognition components presuppose the existence of a previous utterance which the user will refer to when interpreting the system's utterance, i.e. they are restricted to contextual ellipsis. (The various forms of contextual ellipsis which HAM-ANS does not handle at present, such as expansion ellipsis, would not require major structural changes.) Extension to the generation of telegraphic ellipsis, i.e. elliptical utterances which are not linked structurally to any previous utterance, would not be straightforward. In particular, the idea of generating such utterances by deleting constituents of a complete utterance is less natural than it is for contextual ellipsis (cf. Gunter 1963, Betten 1976). On the other hand, if a generation algorithm for telegraphic ellipsis is available which in general produces several candidates, a local anticipation feedback loop can in principle be used to eliminate ambiguous candidates - provided, of course, that the system's ellipsis recognition procedures are capable of processing the candidates generated. Although some of the ellipsis recognition strategies developed to date are applicable to telegraphic ellipsis (e.g. those described by Waltz 1978 and Harris 1980), it should be noted that utterances of this type which are generated by natural language dialogue systems often represent phrases which are specific to the role being taken by the system - e.g. 'Your name, please?' in the hotel situation - and which the system may thus not be equipped to deal with as input.

Not all conceivable ellipsis recognition components would lend themselves equally well to incorporation in a local anticipation feedback loop. First, any relevant knowledge used by the recognition component must be represented in a way which is neutral with regard to whether the elliptical utterance comes from the system or from the user. This is not the case, e.g., with the approach described by Bobrow, Kaplan, Kay, Norman, Thompson, and Winograd (1977), which is restricted to the recognition of the user's answers to the system's questions. Second, the recognition component must operate on the same kind of structure as those returned by the system's ellipsis generation component.This would not be the case, e.g., with Kwasny and Sondheimer's (1981) method, which uses the Augmented Transition Network path of the elliptical utterance.

The anticipation feedback loop described above represents a special case of user modelling in that (a) it is not based on any general assumptions about users beyond the implicit assumption that they interpret elliptical utterances in about the same way as the system itself, and (b) it does not involve the accumulation of specific assumptions about the particular user in the course of the dialogue. The user modelling to be described in the remainder of this paper is more typical in both of these respects.

The Role of the User Model in NP-Generation

Much has been written in AI and related disciplines about the comprehension of anaphoric definite descriptions (for a review see Sidner 1979), much less about their generation (for reviews see Wong 1975 and Heidorn 1977), and almost nothing about the influence of the user model on the cognitive processes underlying their generation - the problem to be investigated in this section of our paper.

In the hotel reservation situation mentioned above, where the system takes the part of a hotel manager who tries to persuade the user to book a particular room, the user model plays a major role in the system's NP-generation process[1]; e.g. when tendentious descriptions of the same object can create any of a variety of impressions of it in the mind of the hearer (as when the same bed is described as 'not too hard' or 'not all that soft') or when the system tries to make the user think that the room has several instances of a desired object when it in fact has only one - two phenomena which will be discussed in more detail below.

The degree of complexity of the NP-generation process is influenced by the type of situational context and by the system's conversational goal. It is useful to distinguish two types of situational contexts and two types of goals on the system's part:

(S1) Both partners are familiar with the details of the particular domain being discussed. For example, in one application, in which HAM-ANS is interfaced to a motion analysis system observing a street crossing, it is assumed that either both partners are looking at the scene or the user is familiar with it (or has a photograph of it).

(S2) The listener must be assumed to have no definite prior knowledge about the particular domain underlying the discourse. For example, in the hotel reservation situation at the beginning of the dialogue, the client, unlike the system, has no definite knowledge about the interior of the room being offered.

(G1) The system has no interest beyond helping the user by responding in a slavishly cooperative way. For example, in the scene analysis application the only task of the system is to give accurate information about the scene observed.

---

1 Although the system HAM-ANS includes a pronominalization component, we must limit the present computational exploration to the pragmatics of *singular definite anaphoric reference* to nonpronominal NPs.

(G2) The system's intention is to influence a decision to be made by the dialogue partner. For example, in the hotel reservation situation the system tries to persuade the client to rent the room.

In this section we will focus on conversational settings which can be characterized by the combination of (S2) and (G2), since the user model plays a much more prominent role there than in situations characterized either by a combination of (S1) and (G1) or a combination of (S2) and (G1). Although most Al work has been devoted to the latter two combinations, we believe that for future Al applications the first combination may become equally important.

The type of conversational setting to which the hotel reservation situation belongs is characterized by the occurrence of definite descriptions in anaphoric chains initiated by sentences containing an indefinite description (e.g. '... next to an antique armchair ... The armchair ...'). We will therefore specify when and how indefinite descriptions are generated by the NP-generation component. To provide a concrete background for the discussion, we first define the task of the NP-generation component in HAM-ANS more precisely and present the technical framework for the proposed computational method.

The Structure of NP-Generation in HAM-ANS

When the verbalization component encounters a logical term in the semantic structure passed to it, the NP-generation component (NP-GEN) is called.

If the term consists of a <quantifier-operator> and a <description> (see Jameson et al. 1980 for details) the complete structure of a definite or indefinite NP has been formed by the answer generation component, so that the only remaining task of NP-GEN is to translate the term into a preliminary NL structure.

If the term contains individual constants (e.g. DESK1) the subcomponent DESCRIBE expands all constants into a term which consists of a <quantifier-operator> and a <description> . Then the NP-generator is called recursively for the translation of the expanded term into a preliminary NL structure as mentioned above.

For example, the term (6) is replaced by the term (7) (as described below) and then translated into (8).

```
(6)   (t-o: AND CHAIR1 BED1)

(7)   (t-o: AND
            (t-s: (q-qt: A)
                  (d-o: AND
                        (lambda: x65 (af-a: ISA x65 CHAIR))
                        (lambda: x63 (af-a: REF x63 COMFORTABLE))))
            (t-s: (q-d: THE (r: 1 1))
                  (lambda: x64 (af-a: ISA x64 BED))))

(8)   ((a (comfortable) chair) and (the bed))
```

It is useful to distinguish two possible readings of the NPs generated, depending on whether or not a particular object in the domain of discourse is being referred to: (a) the *extensional reading,* as in '*The TV set* has remote control' and '*A shower* is available'; and (b) the *intensional reading,* as in '*The TV set with remote control* is standard equipment in our rooms' and 'Unfortunately, *a shower* is not available'.

The semantic structure of NPs for which an intensional reading is intended is constructed during the answer generation phase; it would clearly be impossible to generate constant terms in this case. On the other hand, most NPs which are intended to have an extensional reading are generated by NP-GEN out of constants using the DESCRIBE component. Since in this paper we are focusing on the influence of the user model on the decision process modelled by the DESCRIBE component, we shall examine only the latter case in more detail.

The DESCRIBE component uses various knowledge sources, including referential and conceptual knowledge, inference rules, and a user model. The role of the following parts of the user model in the NP-generation process will be described: existential assumptions, domain-related desires, dimensional preferences, the partner's referential network, a coreference network, and the specific beliefs and desires of the partner.

To make the discussion more concrete, Figures 2 through 4 show simplified versions of some sections of the knowledge sources which are most important for the presentation to follow.

The Generation of Indefinite Descriptions

As long as the system is not trying to confuse or deceive the user by deliberate referential miscommunication, DESCRIBE generates an indefinite NP iff the individual constant is not included in the coreference network, which links the system's referential knowledge to the user model. There are two possible reasons for the presence of such an entry in the coreference network: Either the system has already referred to the object or the existence of the object is implied by the existential assumptions supplied by the user model. In the latter case, the system generates a so-called *pragmatic anaphora*, i.e. an anaphora with no linguistic antecedent. For example, the system may regard as shared knowledge the fact that there are a bed, a door, and a window in the hotel room (see Clark/Marshall 1981). Then, when referring to these objects for the first time it doesn't use an indefinite NP, but rather generates a definite NP as a pragmatic anaphora.

Which existential assumptions are ascribed to the user depends in turn on the user model. For example, if the caller introduces himself as a senior manager, the system will ascribe to the user the existential assumptions associated with luxury hotels, e.g. that the room has a colour TV, so that the system can speak of 'the colour TV' without its having been mentioned previously.

When producing an indefinite description, the system generates a new individual constant, links it in the coreference network to the coreferential object in its referential knowledge, and stores all the information contained in the description in the partner's referential network.

When generating an indefinite NP, DESCRIBE has to look for an appropriate initial characterization, checking the desires ascribed to the partner. For example, in the given context it would not be advisable to introduce CHAIR1 of Figure 2 as 'a broken chair',

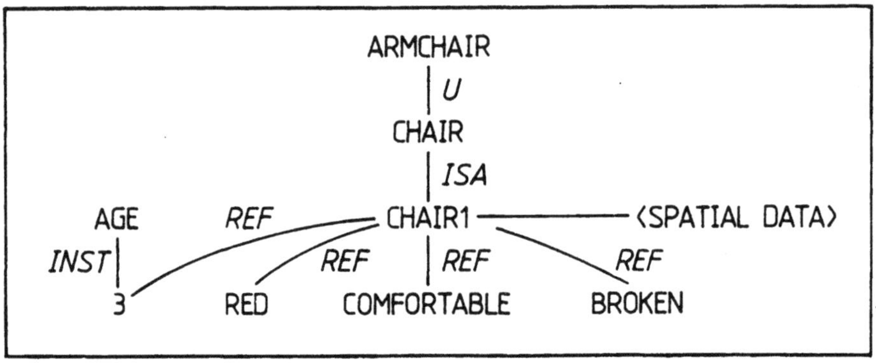

Fig. 2: A small segment of the system's referential and conceptual knowledge.

but on the basis of a preference ascribed to the partner by default 'a comfortable chair' would be a good choice. If there is more specific information in the user model, e.g. that the client has a strong preference for antique furniture, HAM-ANS can describe CHAIR1 as 'a relatively old chair', taking advantage of its ability to generate vague expressions using hedged relative adjectives (see Wahlster 1981). Otherwise, if the client seems to dislike old furniture, the system would prefer the NP 'a quite new chair', as shown in Figure 3. In the current implementation, the amount of information attached to the introducing indefinite NP is controlled by a parameter called VOLUBILITY. In a more advanced

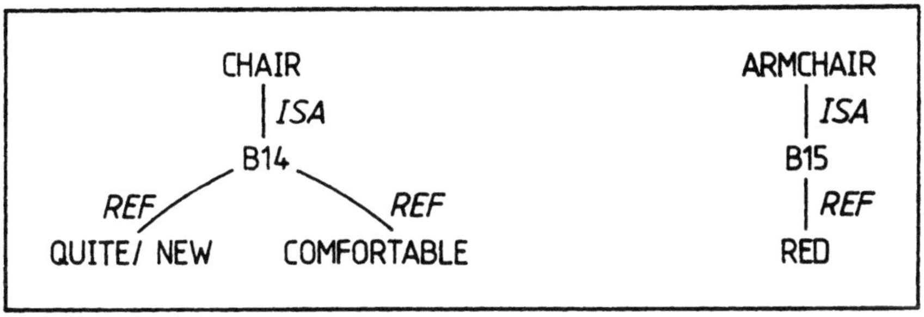

Fig. 3: Two entries in the partner's referential network in the user model.

version of HAM-ANS, the value of this variable should depend, among other things, on a continuous reevaluation of the system's success in creating a favorable impression of the room.

The Generation of Definite Descriptions

Once it has used an indefinite description, the system is entitled to use a definite NP. The process of generating anaphoric NPs is governed by the principle of *confusion avoidance* (Herrmann/Laucht 1976). When the object to be referred to is one of several objects which are stored in the partner's referential network and which belong to the same conceptual class, the system looks among the

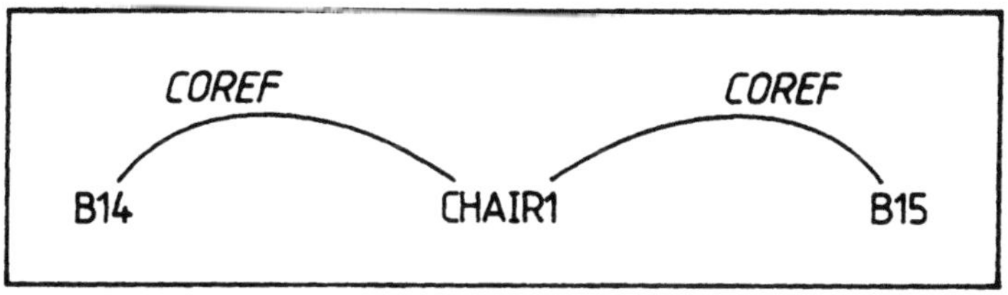

Fig. 4: An entry in the coreference network.

properties stored in the partner's referential net for a subset which distinguishes it from its cohyponyms. Since the search for a uniquely identifying description is independent of the particular role played by the user model, no special algorithm need be developed for this case. (In HAM-ANS the same algorithm is used as in the traffic situation; see Wahlster/Jameson/Hoeppner 1978.)

If there are several properties which refer only to the intended object, we have the case of *multiple codability* studied in psycholinguistics by Herrmann and Laucht (1976). If we omit the possibility of redundant coding from consideration, multiple codability calls for a decision process. HAM-ANS applies the following criteria successively to find a unique property:

* Checking the user model, choose the property which corresponds best to the desires of the partner.

* Choose the property which permits the best subjective discrimination, i.e. the property according to which the objects with which the object in question might be confused differ most strongly.

* Choose the property which has the highest rating on the dimensional preference scale ascribed to the partner (the default preference order in HAM-ANS is simply represented as a list of dimensions, e.g. color, contrast, size, age, ...).

The major new feature of the approach outlined above is that NP-generation in the type of conversational setting under investigation is viewed as an *interest-based* activity which is heavily influenced by the user model. One of the challenging aspects of the approach taken is that the proposed computational methods can also cope with intentional exaggeration and understatement with respect to the cardinality of certain sets of objects.

Suppose, for example, that the client has said 'I trust that there are several chairs in the room' (he plans to hold a business meeting in the hotel room) and HAM-ANS has generated a positive response. In order to reinforce the user's impression that there are enough chairs, the system describes, e.g., CHAIR1 from Figure 2 (which it has previously referred to as 'quite a new comfortable chair') later in the dialogue with an indefinite NP, namely 'a red armchair' (see Figure 3). The system then generates two instances of a chair in the partner's referential net and links them in the coreference network to the same object identifier.

For an illustration of the understatement tactic, suppose the client wishes a single room but the hotel manager is forced to try to rent him a double room. In this situation the hotel manager can try to conceal the real number of beds in the room. After having referred to BED1, the system refers not only to it, but also to BED2 as 'the bed', so that the coreference net contains the assertions (COREF BED1 B1) and (COREF BED2 B1) and B1 is the only bed in the partner's referential world.

It now becomes obvious why the system must construct the coreference network and the partner's reference network and cannot use the simpler approach of marking in the system's referential knowledge the objects and properties which belong to the shared knowledge: In the preceding examples there is no simple correspondence between objects in the system's referential world and those in the partner's. Only by updating the coreference network can the system ensure consistency in intentional exaggeration and understatement.

In this paper we have viewed NP-generation as a process controlled by the user model. It should be clear that there are other factors such as focus (cf. Grosz 1981) and speech act planning (cf. Cohen 1981) which also play a major role in referent identification.

Bibliography

Allen, J.F., 1979, "*A plan-based approach to speech act recognition*".Unpublished PhD Thesis, Dept. of Computer Science: University of Toronto.

Betten, A., 1976, "Ellipsen, Anakoluthe und Parenthesen. Fälle für Grammatik, Stilistik, Sprechakttheorie oder Konversationsanalyse?" In: *Deutsche Sprache,* 4, pp.207-230.

Bobrow, D.G./Kaplan, R.M./Kay, M./Norman, D.A./Thompson, H./ Winograd, T., 1977, "GUS, a frame-driven dialog system." In: *Artificial Intelligence*, 8, pp. 155-173.

Clark, H.H./Marshall, C.R., 1981, "Definite reference and mutual knowledge." In: A.K. Joshi/I. Sag/B. Webber (eds.), *Elements of discourse understanding.* Cambridge University Press, Cambridge, pp. 10-63.

Cohen, P.R., 1978, "*On knowing what to say: Planning speech acts.*" Dept. of Computer Science: University of Toronto, Technical Report, 118.

Cohen, P.R., 1981, "The need for referent identification as a planned action." *Proceedings of the 7th International Joint Conference on Artificial Intelligence.* Vancouver.

Grosz, B.J., 1978, "Ellipsis". In: D.E. Walker (ed.), *Understanding spoken language.* North-Holland, New York, pp. 315-337.

Grosz, B.J., 1981, "Focusing and description in natural language dialogues." In: A.K. Joshi/I. Sag/B.Webber (eds.), *Elements of discourse understanding.* Cambridge University Press, Cambridge, pp. 84-105.

Gunter, R., 1963, "Elliptical sentences in American English." In: *Lingua,* 12, pp. 137-150.

v. Hahn, W./Hoeppner, W./Jameson, A./Wahlster, W., 1980, "The anatomy of the natural language dialogue system HAM-RPM." In: L. Bolc (ed.), *Natural language based computer systems.* Hanser/Macmillan, München/London, pp. 119-253.

Harris, L.R., 1980, "ROBOT: A high performance natural language interface for data base query." In: L. Bolc (ed.), *Natural language based computer systems.* Hanser/Macmillan, München/ London.

Heidorn, G.E., 1977, *A method for generating noun phrases to identify nodes in a semantic network.* IBM T.J. Watson Research Center, Yorktown Heights, N.Y.

Herrmann, T./Laucht, M., 1976, "On multiple verbal codability of objects." In: *Psychological Research,* 38, pp. 355-368.

Hoenkamp, E., 1980, "Spontaneous speech as a feedback process." *Proceedings of the AISB-80 Conference on Artificial Intelligence.* Amsterdam.

Jameson, A./Hoeppner, W./Wahlster, W., 1980, "The natural language system HAM-RPM as a hotel manager: Some representational prerequisites." In: R. Wilhelm (ed.), *GI-10. Jahrestagung, Saarbrücken.* Springer, Berlin, pp. 459-473.

Kwasny, S.C./Sondheimer, N.K., 1981, "Relaxation techniques for parsing grammatically ill-formed input in natural language understanding systems." In: *American Journal of Computational Linguistics,* 7, pp. 99-108.

Rich, E., 1979, "*Building and exploiting user models*". Computer Science Dept.: Carnegie-Mellon Univ., Technical Report, CMU-CS-79-119.

Sidner, C.L., 1979, "*Towards a computational theory of definite anaphora comprehension in English discourse.*" MIT Al Laboratory. Technical Report, TR-537.

Simmons, R./Slocum, J., 1972, "Generating English discourse from semantic networks." In: *Communications of the ACM,* 15, pp. 891-905.

Wahlster, W., 1981, "Implementing fuzziness in dialogue systems." In: B. Rieger (ed.), *Empirical semantics.* Brockmeyer, Bochum, pp. 259-280.

Wahlster, W./Jameson, A./Hoeppner, W., 1978, "Glancing, referring, and explaining in the dialogue system HAM-RPM." In: *American Journal of Computational Linguistics,* Microfiche 77, pp. 53-67.

Waltz, D.L., 1978, "An English language question answering system for a large relational database." In: *Communications of the ACM,* 21, pp. 526-539.

Winograd, T., 1977, "A framework for understanding discourse." In: P.A. Carpenter/M.A. Just (eds.), *Cognitive processes in comprehension.*Erlbaum, Hillsdale, N.J., pp. 63-88.

Wong, H.K.T., 1975, "*Generating English sentences from semantic structures.*" Dept. of Computer Science: Univ. of Toronto. Technical Report, 84.

STATISTICAL MODELING OF THE SPEECH FLOW[1]

Ju. K. Orlov
Institute of Cybernetics
Thilisi, USSR

The ever increasing extent of statistical investigations concerning language, problems of compiling minimum dictionaries, work on machine-produced dictionaries and systems of automatic translations, the analysis of the flow of key words within information systems of different kinds (ASU = avtomatizirovannaja sistema upravlenija = automatic control system, IPS = informacionno-poiskovaja sistema = information-searching system and the like) and computational problems of the structure and extent of data banks (with a given percentage of faults) within such systems lay more and more claim to the theoretical analysis of speech flow with regard to its statistical structure. But a large number of contributions dedicated to relationships of the type "class-frequency" and "size-dictionary" and the like started to be problematic themselves so that a theoretical understanding becomes necessary.

In spite of the fact that problems concerning the quantitative analysis of speech flow are thought to lend themselves to formalization according to their nature (there are no semantic problems, the observable facts directly manifest themselves in numbers), up to the present time there is no strict and complete theory, i.e. a theory that describes and predicts all observable phenomena. What exists is merely a considerable amount of models of different complexity, precision, completeness and validity that describe connections between one or the other kind of quantitative characteristics of speech flow.

Without pretending to give a complete survey, in the present article an attempt is made to carry out a comparative analysis of

---

1 Translated from the Russian by Annemarie Koch-Wiegart.

some models well-known in the corresponding literature, of models that describe and predict the increase of the vocabulary in lexical excerpts and of relationships of the type "class-frequency".

What is the sort of tasks in the practice of statistical investigations concerning language that have to be solved by a theoretical analysis of quantitative relationship that can be observed within lexical excerpts?

1) First of all, the feasibility of a concise description of the results of the statistical calculations, the possibility or reducing most of the figures to only a few (ideally to one figure); it is extremely desirable that these figures allow an immediate linguistic interpretation.

2) The possibility to *compare* with the help of a model description texts of different length and excerpts with one another. This, in particular, is a problem of lexical concentration or of relative richness of the dictionary, when you are to compare the stock of words on excerpts of different lengths.

3) The possibility to go beyond the limited range of the calculation that has been performed - to reconstruct insufficient data and to prognosticate for other lengths (of texts) after regarding the observed data concerning frequency and dictionary.

4) The possibility to confirm one or the other statistical hypothesis concerning data on excerpts. In particular, these are tasks of the type of verification of the hypothesis that two different excerpts are taken from a common lexical complex (examination of the statistical homogeneity of a text or a sequence of key words). This is also the place for tasks of determining a priori - by examining the vocabulary in a text excerpt - which fraction will not be covered in the text excerpt. The last two possibilities are especially important in connection with such applied problems as machine-produced translations and the organization of databanks (the determination of the indispensable size of a machine

dictionary or data banks with a given percentage of faults).

All models that try to solve the enumerated tasks can be categorized into two basic types. In the first place there are empirically selected analytical relationships that allow to describe the observed results in a concise way and that are not based on any hypotheses about the "organization" of a text; in the following we are going to call them descriptive models. In the second place there are models that are based on one or the other theoretical assumption about the structure of a text, about a set of probabilities for the occurrence of words contained in the text. We are going to call such models structural. Their value is undoubtedly higher than the value of the descriptive models as only the former have the possibilities 2) - 4); besides, even in such cases, where a structural model describes reality inadequately, the negative statement that it is inadequate does not simply lead to the conclusion "the formula does not work", but permits to draw a pithy conclusion about the incorrectness of the ideas that the model is based on (further on we are going to demonstrate this for one of the models).

The general platform of ideas of all structural models is Herdan's concept (Herdan, 1966) that any speech utterance can be regarded as a random sample from some invariable lexical whole, every word of which is characterized by an invariable probability of its use in speech (within a text). Up to the present day there has been sufficient evidence that - strictly speaking - this concept is false: the failure to produce frequency dictionaries concerning languages in general that forced the researchers to deal with increasingly narrow "sublanguages", statistically relevant changes of frequent words in subexcerpts even when taken from one text (Tokarev, Jakubaitis 1969; Bektaev, 1978), the phenomenon of the concentration of rare words in local groups according to the length of the text (Boroda et al., 1977; Nadarejsvili et al. 1977) and others. Only this concept, however, allows to use combinatorial logic, probability theory and mathematical statistics. This makes evident that structural models based on such grounds will always be in discord with reality to a certain degree. Fortunately such discord is not immense and for the majority of applied tasks it is within well acceptable limits.

Let us take a closer look at the basic quantitative consequences of Herdan's concept. V.M. Kalinin's contributions contain a strict conclusion about them (Kalinin, 1964; 1965); there you also find a very concise criticism of Herdan's concept (cf. also Muller's paper (Ch. Muller, 1976)).

Let us consider a lexical excerpt of the size of N_o words for that the size of the dictionary v (N) and the lexical frequency spectrum $\{ v_m(N_o), m=1,2.. \}$ are known, where $v_m(N_o)$ denotes the number of different words that occur in this excerpt m times. It is obvious that

$$v(N_0) = \sum_m v_m(N_0)$$

What will the size of the dictionary and the lexical frequency spectrum be like for a subexcerpt of size N $(N < N_o)$ taken from our excerpt, if it is sampled without replacement. For the sake of simplifying the discussion, we visualize an excerpt of size N_o as an interval on a linear axis

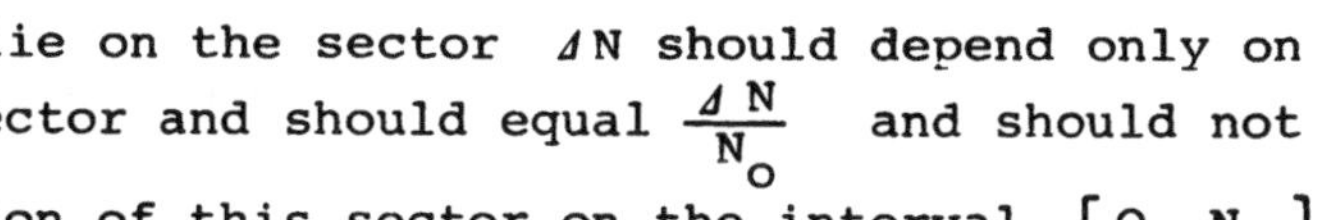

and the occurence of a word as a point on this graph. Herdan's concept corresponds to the claim that the points should be equally distributed on the axis and that the probability for a certain point to lie on the sector ΔN should depend only on the length of this sector and should equal $\frac{\Delta N}{N_o}$ and should not depend on the location of this sector on the interval $[0, N_o]$. A mathematician would object that we replaced the binominal distribution, which is significant for a random sample of words, with the Poisson distribution, which is significant for the number of points that are part of the sector ΔN. This becomes possible as the general number of words used in the actually interesting cases is high, but the frequency of each word in particular is not high; under these circumstances the Poisson distribution does - practically speaking - not differ from the binomial one (a well-known theorem of probability theory). We will find the mathematical expectation for the number of different words at least one of which will be on the section $[0, N]$; to put it differently, we will get the expected size of the dictionary in the subexcerpt of size N. It need not be underlined that the dictionary of the subexcerpt will be less in size than the dictionary of the whole excerpt of size N_o. The easiest way is to calculate how much less the dictionary is represented in the subexcerpt. In the first place, it does not contain words that occur only once (i.e. words you come across only once in the total number N_o) and that are part of section $[N, N_o]$. The probability for each such word to belong to this section equals

$$\frac{N_0 - N}{N_0} = 1 - \frac{N}{N_0}$$

As the total number of such words is $v_1 (N_o)$, the expectation for the number of those that are part of section $[N, N_o]$ equals

$$\left(1 - \frac{N}{N_0}\right) v_1 (N_0)$$

In the second place, the interval $[0, N]$ does not contain those words that you found exactly twice within the total number N_o and that both were part of $[N, N_o]$. The probability for each one to lie on this section is as before $1 - \frac{N}{N_o}$; the probability to get there in both cases equals the square of this probability, i.e. $\left(1 - \frac{N}{N_o}\right)^2$. When we multiply this quantity by the total number of words occurring twice $v_2(N_o)$, we will get the expectation for the number of words occurring twice that are not part of $[0, N]$:

$$\left(1-\frac{N}{N_0}\right)^2 v_2(N_0)$$

If we proceed analogously, we obtain the expectation for the number of words occurring m times (within N_o) with all occurrences belonging to $[N, N_o]$ (or, which means the same, none of which is part of $[0, N]$):

$$\left(1-\frac{N}{N_0}\right)^m v_m(N_0)$$

To obtain the expected dictionary size for N, it is obviously necessary to subtract the expected number of words not belonging to the interval $[0, N]$ from the total number of words in the dictionary for N_o:

$$Ev(N)=\hat{v}(N_0)-\hat{v}_1(N_0)\left(1-\frac{N}{N_0}\right)-v_2(N_0)\left(1-\frac{N}{N_0}\right)^2-\dots-\hat{v}_m(N_0)\left(1-\frac{N}{N_0}\right)^m-\dots$$

or

(1) $$Ev(N)=v(N_0)-\sum_{m=1}\left(1-\frac{N}{N_0}\right)^m \hat{v}_m(N_0),$$

where E stands for the mathematical expectation, and ^ marks an empirically observed quantity.

The derived expression represents (if we are going to regard N as a continuous variable) a curve of expected values of the vocabulary on a random subexcerpt of size N taken from an excerpt of size N_o. It allows to solve such tasks as the comparison of relative lexical richness of two excerpts (or texts) of different size (it suffices to calculate the expected vocabulary according to the frequency data of the two excerpts for one and the same value (for details, cf. supplement II to Nadarejšvili et al., 1977)). Considerable deviations of the real data from the results obtained by calculation according to formula (1) testify that the subexcerpt of size N should not be regarded as a random product. In this way formula (1) may serve as an indicator of distinction between a coherent text and a random text; further on we will return to this question.

Analogous reasoning that we will omit leads to the following expression for the expectation values of a lexical frequency spectrum for N if the spectrum for N_o is known:

(2) $$Ev_m(N) = \frac{1}{m!}\left(\frac{N}{N_0}\right)^m \sum_{j=m} \frac{j!}{(j+m)!}\left(1-\frac{N}{N_0}\right)^{j-m} \hat{v}_j(N_0)$$

It is essential for our discussion that $N < N_o$; V.M. Kalinin, however, reached at formulas (1) and (2) in his above mentioned publications after some more general considerations and showed that a replacement of the actual values of the dictionary size and spectrum for N_o with their expectation values on the right hand side would remove the limitation for N and it could be either less or greater than N_o. To put it differently, knowledge of the expectation values of the frequency spectrum for some excerpt represents an equally complete information about a text as does knowledge of the complete set of word probabilities $\hat{v}_m$.

The curve of increase of the expected dictionary size is a continuous curve (thanks to the use of the Poisson approximation) which allows to carry out a normal differentiation. Therefore it is possible to combine the problems of the frequency spectrum and of the increase in dictionary size by means of differential equations that have also been developed by V.M. Kalinin:

(3) $$\frac{d^m}{dN^m} E\,v(N) = (-1)^{m+1} \frac{m!}{N^m} E\,v_m(N); \quad m = 1, 2, \ldots$$

Thus it becomes evident that the knowledge of at least one of the components determines the other one. It is e.g. quite clear that a "well-behaved" curve of increase in dictionary size (even an empirical one) must at least have a large number of non-vanishing derivatives, as the number of non-zero terms in the spectrum is large in usual lexical excerpts. The exact way of generally judging the quality of the model is to do a complex verification; if a curve of increase in dictionary size suggests itself, then it is necessary to calculate the frequency spectrum using equation (3) and to check to which degree it coincides with actual observations. And vice versa, any kind of approximation for the spectrum, for frequency curves "class-frequency" and the like, have to be transferrable to curves of increase in dictionary size (this is better realized with formula (1)). And if one part can be well described but the other one cannot, then this is an obvious indication for the inadequacy of the model which makes it useless for prognostic purposes. Such a model is only capable of fulfilling descriptive functions.

Equations (1 ÷ 3) play the role of a kind of "grammatical rules" that the models for the increase in dictionary size and for the lexical frequency structure are subject to. This is of course far from being a complete list of such rules that can be inferred and are actually inferred (cf. Kalinin 1964; 1965, as well as Krutikova, 1968) from Herdan's concept, but their evaluation would lead us astray from the basic subject of the present work.

Starting from the general considerations that we have introduced above, let us investigate some models of statistical text organization that are well-known in the literature relevant to the subject; let us begin with the descriptive models. All of them came into being because of one or the other requirement of language statistics, and the very first models are characterized by a most "straight-forward" approach towards the answer to interesting questions.

In particular, the necessity to compare two texts of different lengths by means of the relative (but not absolute) lexical variety lead to the introduction of the Type-Token Ratio (abbr. TTR) that has been occurring in the relevant literature up to now and that stands for the ratio of the dictionary size of a text and its length $\frac{v(N)}{N}$. Apparently it was hypothetically assumed that this ratio must be approximately constant for different excerpts of the same text in the same way as the relative frequency of any word is roughly constant. By defining:

$$\frac{v(N)}{N} = T = \text{const.},$$

we get:

$$v(N) = TN, \tag{4}$$

i.e. this consideration implies a linear increase of dictionary size with increasing excerpt size N. According to (3) the frequency spectrum of such a text should list only words occurring once, because function (4) has just one non-vanishing derivative. Such a "text" cannot emerge as the result of a random excerpt from any totality you like, because the exclusion of repeated word use makes it a non-random excerpt.

In practice, of course, it was immediately realized that the TTR decreases considerably fast with increasing N within one and the same text; this index, however, has continued to be used in literature up to now (its use is accompanied by remarks that text excerpts of the same length should be compared; but in that case it would suffice to simply compare the dictionary sizes of these excerpts). Of all quantitative lexical models known to the author this is probably the only one that is absolutely useless. For the same purpose, i.e. as a measure of the relative lexical variety, G. Herdan proposed the following ratio of the logarithm of the dictionary size and the logarithm of the text size:

$$\nu = \frac{\log v(N)}{\log N}$$

At first sight this ratio appears to be more constant within the limits of one text. In case of ν being strictly constant, the dictionary size becomes a power function of the excerpt size:

$$v(N) = N^{\nu} \quad (5)$$

This function is differentiable to all orders, and its graph on a log-log scale (log v, log N) is a straight line through the origin (fig. 1). On this scale, the actual increase in dictionary size is represented by a graph of small convex curvature. That is why the straight line and the curve intersect with an angle depending on N, and the prognosis concerning the dictionary size is bad especially for large N when based on the value for v calculated from an excerpt of a certain size.

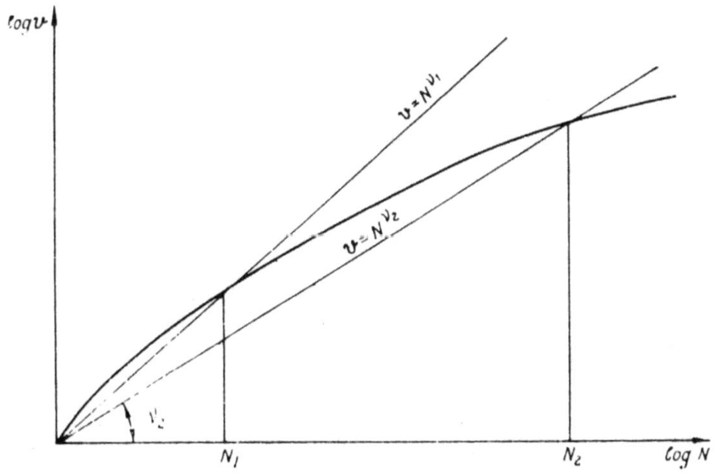

Fig. 1

Herdan's ratio is, however, applicable for the comparison of different excerpts, if their difference in size is not too large and the excerpts themselves are quite similar (for the purpose of orientation about 10,000 word occurrences may be regarded as an approximate upper limit for applicability).

For large sizes (in the order of ten or a hundred thousand word occurrences) Giro's ratio is more suitable

$$R = \frac{v(N)}{\sqrt{N}}$$

Within one text this is in fact a slowly varying function (for large sizes) and can be used not only as an approximate measure of lexical concentration (of relative lexical variety) but also for obtaining an estimate for the dictionary size:

$$v(N) = R\sqrt{N} = RN^{0.5} \tag{6}$$

On a log-log scale, the graph of this type of function is also a straight line but with a constant slope of 0.5. This is a fairly "typical" slope of curves of increase in dictionary size (later on when examining more complete models we shall see why this is so) for the most common lengths of texts and excerpts. The first sections of a text are badly described by function (6) (cf. fig. 2) and there Giro's ratio does not remain constant.

The inspection of figures 1 and 2 reveals the fundamental inadequacy of curves with constant (i.e. constant with respect to N) exponents for the increase in dictionary size. All these curves appear as straight lines on the log-log scale, but the actual increase in dictionary size is represented by a graph of convex curvature. And although the curvature is small, in general a good approximation by a straight line can only be achieved on limited sections.

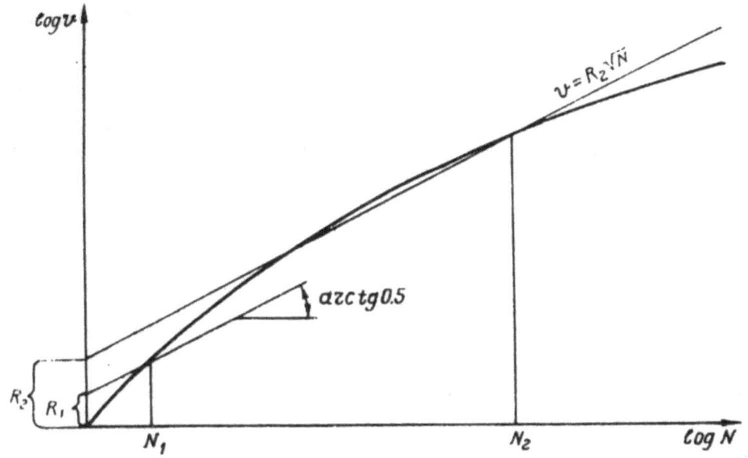

Fig. 2

Table 1

N	$\hat{v}(N)$	$\frac{v(N)}{N}$	$R = \frac{\hat{v}(N)}{\sqrt{N}}$	$\gamma = \frac{\ln \hat{v}}{\ln N}$	$v(N) = R\sqrt{N}$	δ, %	$v(N) = N^{\gamma}$	δ, %
10000	2688	0,269	26,88	0,859	3110	+15,8	2150	—20,0
20000	4249	0,213	30,00	0,844	4100	+3,56	2740	—35,5
40000	6193	0,155	31,00	0,824	6220	+0,44	4750	—29,4
60000	7629	0,128	31,40	0,813	7620	—0,12	6550	14,1
100000	9834	0,098	31,10	0,799	9830*	—	9840*	—
200000	13389	0,067	29,90	0,779	13920	+3,97	17300	+29,2
300000	16104	0,054	29,40	0,769	17080	+6,06	23900	+48,3
Initial data for prognosis					$N = 100000$, $\hat{v}(N) = 9834$			
Selected parameter values					$R = 31,10$		$\gamma = 0,799$	

*The asterisk denotes theoretical values of the dictionary size that coincide with the actual ones "by construction".

Table 1 shows investigations of the ratios of the increase in dictionary size for the example of Latvian newspaper texts (Latviešu ..., 1969). It is evident that for the example given here Giro's ratio works much better than Herdan's ratio with respect to the possibility of prognosing for the dictionary size. In both cases the values of the constants (v and R) were determined at an excerpt size of N = 100.000. A comparison of figures 1 and 2, however, shows that for small N the reverse picture can be observed - Herdan's ratio (curve (5), to be more precise) is a better approximation to reality than curve (6).

The models we have been investigating contain just one fitting parameter that completely determines the applicability of the model. Such models with just one parameter are convenient because their application requires minimum initial information - it is sufficient to have a single observation of the vocabulary in an excerpt. It is this circumstance together with the case of calculation that assured these models of a broad reputation.

Let us now pass on to more complicated descriptive models - models with two parameters.

Ju. A. Tuldava (1974) offers the following relation for the description of the dictionary size

$$(7) \qquad v(N) = \frac{aN}{b + N} \; ; \quad a, b \text{ — const.}$$

This formula, in fact, permits a good approximation for limited sections of the increase in dictionary size. With increasing N function (7) approaches the value a asymptotically, that is why Ju. A. Tuldava proposes to regard this function as an indicator of the "potential lexical variety of a given text". One cannot, however, agree with this suggestion, because the frequency structure of a "text" for which the increase in dictionary size is described by formula (7) is certainly not at all similar to the frequency structure of actual texts.

When differentiating function (7) repeatedly, we find the expression for its mth derivative:

$$\frac{d^m}{dN^m} Ev(N) = (-1)^{m-1} \frac{m!ab}{(b+N)^{m+1}}$$

Substituting this in (3) and solving for $v_m(N)$ yields the formula for the expectation values of the frequency spectrum for N:

$$Ev_m(N) = \frac{ab}{b+N}\left(\frac{N}{b+N}\right)^m$$

This expression clearly shows that in this case the frequency spectrum obeys an exponential law - the number of different words occurring m times decreases with increasing m in geometrical progression. As we will see later on from the discussion of models that deal with this question, for actual frequency structures the dependence of $v_m(N)$ on m is similar to a power law. Practical examinations show that only the first two or three members of the frequency spectrum coincide satisfactorily with the results of theoretical calculations. This means that only on limited sections curve (7) coincides with the "actual" curve of increase in dictionary size as far as the first two or three derivatives are concerned; further considerable discrepancies begin. The sections of the curve where these discrepancies appear move the further away from the origin the higher the order of the derivative. It results from this that formula (7) is capable of a good approximation only for limited sections of the curve representing the increase in dictionary size, and its limiting value a is a "potential dictionary" of a "text", the frequency distribution of which is given by an exponential law. One thing, however, needs mentioning: the constant a is correlated with the exponent of the relative lexical variety because the steeper the slope of the actual curve of increase in dictionary size, the higher the limiting value a of the corresponding approximated function (7). It is true that the level of uniformity of the text influences the behavior of this constant (in coherent texts with a strong tendency to form clusters the curve of increase in dictionary size will have a greater (positive) slope, which leads to a greater value of the constant a so that a unique linguistic interpretation of this constant will hardly be possible).

In order to conclude the question about formula (7), it is necessary to underline that this relation being extremely simple from an analytical point of view is actually very suitable to describe limited sections of the increase in dictionary size and is especially convenient (as are, by the way, all formulas with two parameters) for the purpose of interpolation - for estimating the increase in dictionary size between two observed points. The prognosis concerning the dictionary size for both the "forward" and "backward" directions from the observed section for arbitrarily large distances with this formula is completely doubtful.

V.V. Nešitoj (1977) suggested a formula for the increase in dictionary size

(8) $$v(N) = N^{1/\sqrt{1 + a \ln N}}$$

that looks like a one-parameter formula with the single parameter a. This formula was obtained as an approximate corollary of the hypothesis that the word probabilities in the general lexical totality are distributed according to Vejbulla's law (so that the origin of Nešitoj's model is a theoretical hypothesis about the statistical structure of a text). A comparison with the empirical material, however, illustrated that a correction was necessary, namely to assume that a is a linear function of lnN:

(8a) $$a = a_o + k \ln N$$

Thus we have to regard this formula as a descriptive one with two parameters a_o and k that are not in any way connected with the original (in the construction of V.V. Nešitoj) distribution of Vejbulla. Its theoretical analysis is very difficult because the expressions obtained for the frequency spectrum are extremely cumbersome and their complexity increases considerably for consecutive values of m. Practically speaking, the formula (as, by the way, formula (7) as well) badly describes the first section of the text, but for medium and large sizes (according to V.V. Nešitoj $N \approx 10^4 \div 10^8$) with both parameters fitted a fairly exact description is possible. In contradistinction to (7), (8) is suitable for quite exact prognostications for both "forward" and "backward" directions from an observed section of increase in dictionary size, describing not only coherent but also random ("well-mixed") texts equally well. Calculations along this line are comparatively simple; the parameters a_o and k are conveniently found by a least square fit to a large number of observed points.

The survey of similar models could be continued but this would not make sense. A large number of comparatively simple analytical expressions can be suggested for the approximation of the increase in dictionary size. An increasing number of parameters means attaining better and better approximations (it is true though that the price for this is an extension of the informational model that is necessary for getting started), but at the same time the "transparence" of the models is lost, i.e. the possibility of a linguistic interpretation of their parameters. Interesting results can in fact only be obtained with structural models.

However, before going on to discuss structural models let us investigate descriptive models concerning the lexical frequency structure.

One such model is the following relation well-known as Zipf's law:

(9) $$p_i = \frac{k}{i} \; ; \; i = 1, 2, \ldots$$

where p_i denotes the frequency of the i-th word in a particular dictionary (arranged in the order of decreasing frequency) and k is a constant close to 0.1.

This relation looks like a straight line on a log-log scale (log i, log p_i) with an angle of 45° to the axis. And indeed nearly on every frequency graph you will find an almost rectilinear section and some of them are rectilinear throughout. It is true though that on the lower parts of the graph there are always those characteristic "steps" caused by rare words - by those occurring once, twice and so on. Attempts to make Zipf's relation more flexible resulted in Kondon's formula

(10) $$p_i = \frac{k}{i} \; ; \; k, \gamma \text{ — const.}$$

and Mandel'brot's formula

(11) $$P_i = \frac{k}{(B + i)^{\gamma}} \; ; \; k, \gamma, B \text{ — const.}$$

In the latter formula the first section of the curve is "rounded off" (because of the constant B), with both formulas the rectilinear part may be "rotated" with respect to the axes and the slope is given by the constant γ.

(11) was theoretically developed as a consequence of the process of optimum coding (Mandel'brot, 1957), but neither this formula nor (10) could describe or account for the formation of these "steps" of rare words and the curvature of this part of the curve. This curvature is so immense that some authors porpose to determine the constant γ separately for different sections of the frequency graph; a more radical approach would be to choose different functions for the description: the arc of a circle (Piotrovskij, 1975) or curves of second order (Alekseev, Byčkov, 1977) (on a log-log scale, of course) and so on.

The general disadvantage of all these constructions is the lack of any prognostic power. Even if it is possible to find parameters describing the frequency distribution of some excerpts, it remains completely unknown, what the curve would look like for another excerpt of the same text. In general, also the linguistic meaning of these parameters remains vague.

Apart from the general form of the frequency curve we are also interested in the distribution of step lengths, i.e. in the numbers of words occurring once or twice as well as in other components of the frequency spectrum. Yule who obviously tackled this question first suggested the following formula (Yule, 1944):

$$(12) \qquad v_m(N) = \frac{\text{const.}}{(m - a)\ (m - a - 1)}$$

where a is a small constant. The "staircase" calculated according to this formula is formed like an arc. For a = 0 it becomes straight and may become a continuation of Zipf's straight line. Booth (Booth, 1967) obtained the same result and combined the "staircase" of rare words and Zipf's straight line (cf. also Arapov, Efimova, 1975). A certain move forward in this direction was achieved by publications of the author of this contribution (Orlov, 1969; 1970; 1976). By finding a condition for normalization, he succeeded in relating all constants that are contained in the formula for the curve (11) as well as in the (discrete) frequency spectrum (12). If for an excerpt of size Z the word occurrences are precisely described by the Zipf-Mandel'brot law (in the version (11)), we get for the sequence of relative frequencies:

$$(13) \qquad p_i = \frac{k}{B + i} ;$$

$$k = \frac{1}{\ln(Zp_a)} ; \quad B = \frac{k}{P_1} - 1 :$$

where p_1 stands for the relative frequency of the word that occurs most often (We limit our discussion to the case $\gamma = 1$; in Orlov, 1976, the other special cases can be found.). The sum over the first v(z) numbers p_i equals unity, where

(14) $$v(z) = \frac{Z}{\ln(Zp_1)} - B = Zk_j - B$$

represents the dictionary size of the excerpt. The "staircase" of rare words (frequency spectrum) is given by

(15) $$v_m(Z) = \frac{v(Z)}{m(m + 1)}$$

The "steps" described by this formula may be constructed to continue the straight line (13); in other words, (13) turns out to be an upper envelope of these steps. (The existence of the steps and the fact that they are situated below the straight line (13) have to be accounted for in (13) in the normalization condition for the first v(z) terms of the sum.)

The only formal reason, why (13) is taken to be the upper envelope of the "staircase" rather than to run through the centers of the "steps", is the fact that in this way the formulas assume the simplest form (in awkward cases it is necessary - as shown by M.V. Arapov (Arapov, Efimova, 1975) - to introduce an additional constant determining the "coefficient of division" for the "steps"). In principle the problem is devoid of real meaning, because the Zipf-Mandel'brot law taken as the basis of all such constructions is not fulfilled in the general case anyway and the "staircase" strongly deviates from the continuation of the straight line found in the medium range of frequencies.

However, relations (13) - (15) offer a new possibility that was not available in earlier investigations of the lexical frequency structure. It is easy to see that none of the three specially defined parameters in Mandel'brot's formula (11) (of rather vague

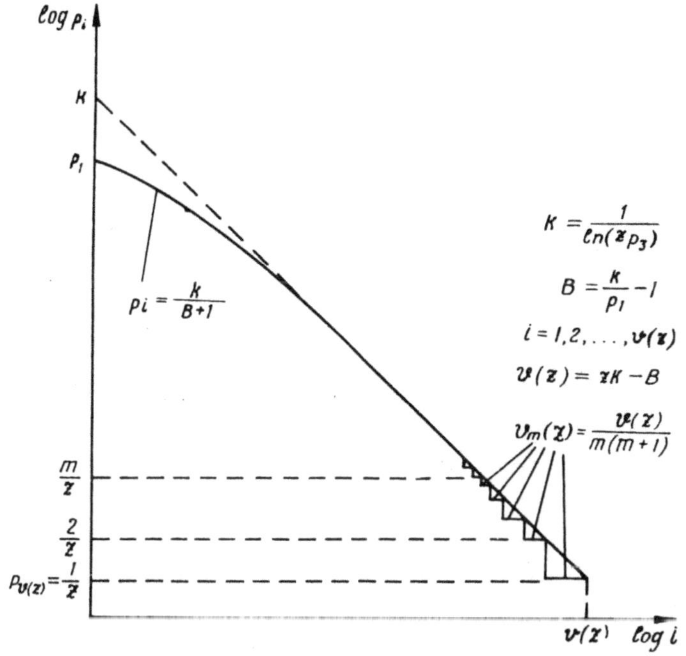

Fig. 3

linguistic meaning, by the way) is left in the formulas (13)-(15) so that z and p_1 become immediately accessible constants of the excerpt (i.e. its size and the frequency of the most frequent word). These "rigid" formulas (a model without fitting parameters) permit to solve problems of *classification* in a very simple way. If we compare the actual vocabulary of a text excerpt with calculations after (14), the degree of deviation enables us to judge how well the Zipf-Mandel'brot law is fulfilled by the text excerpt. This is done in the way shown in fig. 3. In (14), z is identified with the actual text size and p, with the actual frequency of the most frequent word. Examinations using (13) and (15) have shown that the agreement (13) and (15) is good over the entire frequency range of the excerpt if the dictionary size is close to its theoretical value v(z).

The result of the classification carried out proved to be extremely astonishing (Orlov 1970, 1974; Nedarejšvili, Orlov, 1971). It was found that (13) - (15) give an excellent description of the frequency data of some important literary texts. The deviations of the dictionary sizes from the theoretical values according to (14) were smaller than ± 20%. Texts, however, that represent excerpts or collections of many texts showed deviations that were twice as large; excerpts were found to have an increased dictionary size, whereas large collections of many texts had a reduced dictionary size.

A great number of examples of this phenomenon can be found in Orlov (1970, 1978); here only such data is presented (table 2) that has not been published previously. Deviations are marked with an asterisk, if the text was not a complete literary work (of more than 10,000 words). It is easily realized that for single texts the deviations of the actual dictionary size from the theoretical value after the Zipf-Mandel'brot law do not surpass 11%. Later on we will return to the discussion of table 2.

As we do not intend to discuss some details of this phenomenon that have been mentioned in other papers (non-applicability of the Zipf-Mandel'brot law in short contemporary texts, its applicability in various cases of large excerpts (volumes, books) of very long texts), we will only mention that the Zipf-Mandel'brot law cannot be fulfilled by complete texts unless the text is produced in such a way that its entire future length is taken into account from the start. As can easily be inferred from (13), the constants k and B decrease with increasing z (corresponding to the entire length of the text in this case), the frequency curve becomes steeper (for the same value of p_1). Reversely, a decrease of z leads to an increase of k and B, the frequency curve drops less drastically, the fraction of frequent words is raised. Figuratively speaking, the more you want to tell, the less you should repeat yourself and vice versa. The Zipf-Mandel'brot law in its canonical form (without cut-off in the tail of the distribution) turns out to be a *standard of repetitions* for a literary work, but not at all a statistical law for language in general.

Table 2

Excerpt and Source	N	p_1	v(N)	v(N) cf.(14)	%
Dante. "La Divina Commedia"	1o1,554	o.o39	13,oo4	12,25o	+5.8
Dante. "La Divina Commedia", "Inferno"	34,126	o.o41	6,598	4,7oo	+28.8 *
Dante. "La Divina Commedia", "Purgatorio"	34,o42	o.o39	6,45o	4,73o	+26.7 *
Dante. "La Divina Commedia", "Paradiso"	33,386	o.o39	6,273	4,75o	+24.3 *
Dante. "La Vita Nuova"	14,392	o.o45	2,275	2,225	+2.2
Mickiewicz. "Pan Tadeusz"	64,51o	o.o32	9,25o	8,48o	+8.3
Mickiewicz. "Pan Tadeusz" books I-VI	34,28o	o.o32	6,823	4,9oo	+28.2 *
Mickiewicz. "Pan Tadeusz" book I	6,587	o.o37	2,257	1,195	+47.5 *
Shakespeare. "King Lear"	25,471	o.o39	3,391	3,71o	-9.4
Byron. "Don Juan"	13o,745	o.o46	14,411	14,9oo	-3.5
Lermontov. "Geroj našego vremeni"	4o,363	o.o33	5,925	5,95o	-o.4
Pisarev. "Realisty"	48,354	o.o54	6,348	6,15o	+3.2
Paustovskij. "Dym otečestva"	57,ooo	o.o34	7,829	7,53o	+3.8
Gladkov. "Povest' o detstve"	127,917	o.o6o	12,821	14,25o	-1o.o
Camus. "L'Etranger"	14,33o	o.o4o	2,535	2,255	+11.o
Frequency Dictionary of the Russian Language	1,o56,382	o.o41	39,268	99,ooo	-6o.o *
Frequency Dictionary of the English Language	1,o14,232	o.o69	5o,4o6	91,ooo	-44.5 *

But apart from this phenomenon that permits to predict the frequency structure and the dictionary size of an important literary work (provided that its length and the frequency of the most frequent word are known), the descriptive models of the lexical frequency structure were not found to possess prognostic power. This is associated with their general static nature. But in order to construct dynamic models more profound hypotheses are needed.

From a practical point of view in language statistics structural models are also more complicated, more extensive and less exact (as far as their descriptive value is concerned) than the better descriptive models. That is why descriptive models are so popular in literature while structural models are mainly applied by their authors. It is only structural models that are capable of a complex description of statistical text phenomena as they describe not only the increase in dictionary size but also the dynamics of the frequency structure and allow to introduce some degree of certainly into prognoses. Usually even the meaning of deviations from reality within such models is comprehensible; such deviations make possible a better understanding of the nature of the quantitative "construction" of the speech flow. The constants that enter these models may usually get an immediate linguistic interpretation. Descriptive models are less adequate for a clear interpretation; the more exact such a model is the more correction parameters have to be introduced whose linguistic meaning becomes all the more vague.

Strictly speaking the simplest model of the structural type can be found in V.M. Kalinin's formulas (1) and (2) that are based on Herdan's concept. Not only does this model correspond to reality and possess prognostic power (at least as far as prognoses "backward" are concerned; prognoses "forward" are only possible for the double size $N \leq 2N_0$), discrepancies between prediction and reality make evident that the subexcerpt the data of which did not coincide with the model prognosis is no random excerpt of the totality that the excerpt was taken from whose data formed the basis of the prognosis. This makes V.M. Kalinin's formula (1) a standard of special kind for the verification of the hypothesis on the homogeneity of a given text (or of a body of texts).

Formula (1) describes the increase in dictionary size for an ideal random text. Such a "text" can be achieved for example by carefully mixing a text that is written on a number of cards. When arranging them in a random order the corresponding curve of increase in dictionary size should exactly coincide with the calculations after formula (1); minor deviations may occur due to random fluctuations. V.M. Kalinin calculates (in Kalinin, 1965) the dispersion of this process that can be taken as the basis for calculations of confidence intervals for statistical assessment (cf. Orlov, 1978).

It is interesting to know that quite often in the practice of language statistics procedures are arranged to approach in one way or the other - according to their authors' concept - the mixing described above. Of such kind are, in particular, examinations of the vocabulary in excerpts of different size of an evidently heterogeneous original material (for example of newspaper language), as long as the texts of the different columns are presented in the same proportion in the subexcerpts as in the whole body of material ("representative" or "proportional" subexcerpts). Of such kind are also the so-called "mechanical" subexcerpts that are formed by splitting off equally large fragments of equal spacing from a coherent text (for example the upper third of a page). Actually, such procedures are completely unnecessary as their results can be exactly predicted by formula (1). If, however, deviation should still occur, this would simply show that the applied procedure of "averaging" the subexcerpts was not successful and the corresponding subexcerpt is not "representative"; at the same time the degree of deviation allows to assess the coherence of the text sections that make up the subexcerpt.

In spite of all its advantages "Herdan's model" (1,2) suffers from two serious drawbacks. First it does not allow somewhat reliable prognoses "forward" and secondly it requires rather detailed information about the original text before getting started - the actually observed frequency spectrum and dictionary size. Therefore such questions as the "structure" of the frequency spectrum itself cannot be solved by this model.

G. Herdan's concept assumes that the frequency spectra that are to be observed are "created" by a set of word probabilities in the general lexical totality. In the Poisson approximation that we made use of when deriving formula (1), the relation between this set and the expectation of the dictionary size and the spectrum are written down in the following manner:

(16) $$\Sigma v(N) = \sum_{i=1}^{v} (1 - e^{-Np_i})$$

(17) $$\Sigma v_m(N) = \sum_{i=1}^{v} \frac{(Np_i)^m}{m!} e^{-Np_i}$$

where p_1, p_2, ..., p_v stand for the probabilities of all v different words that the general totality consists of.

The main problem is that we neither know the set $\{p_i\}$ nor the expectation values E v(N) and E v_m(N) but only some arbitrary actual realizations v(N) and v_m(N) and the transition to the set $\{p_i\}$ by taking the inverse of the formulas (16, 17) is impossible. Therefore everyone who wants to construct a probability model of the text will either have to try to guess the probability set $\{p_i\}$ "blindfold" (as for example V.V. Nešitoj tried to do) or to circumvent the problem somehow.

In the papers mentioned above (Kalinin 1964, 1965) V.M. Kalinin constructed a probability model assuming that the word probabilities are distributed according to Zipf's law:

$$p_i = \frac{k}{i}; \quad i = 1, 2, \dots, \bar{v}$$

$$\sum_{i=1}^{v} p_i = 1$$

By inserting the obtained values p_1 into the formulas (16, 17) he calculated the curve of increase in dictionary size and the frequency spectra. When he contrasted the obtained analytical relations with the actual observations afterwards, he tried to achieve optimum agreement with the observations by a suitable choice of the parameters k and v of the model and thereby to determine the values of these parameters for the text in question. This is what he writes himself about the results of his efforts "... the constant v strongly depends on the values m and N for which it was determined and partly loses its significance for linguistics if it turns into a conditional working parameter that records deviations of the true probabilities p_i from Zipf's law."

In other words: the hypothesis that the word probabilities of language are distributed according to Zipf's law has not been confirmed.[1] This result is important on principle. Up to now it has been assumed that it is sifficient to smooth the observed frequency set in some way or the other to obtain the corresponding set of probabilities but this procedure only makes sense if all frequencies were reliably determined (practically speaking, if the smallest observed frequency has a value of greater than ten). In the case of lexical excerpts this is an illusion anyway - in any excerpt words occurring once represent a considerable part of the vocabulary and the observed dictionary size is much lower than one would have expected from the increasing size of the excerpt. This general insufficiency of lexical excerpts is an important obstacle to their statistical examination. This is exactly why as J. Carroll (Carroll, 1968) remarks "estimates for probabilities of a theoretical general totality that the excerpt is taken from are systematically shifted to higher values if the frequencies

1 A similar model that assumes that the word probabilities of language are distributed according to Mandel'brot's law (formula (11)) is applied by E. Vorončak for measuring the lexical concentration ("Methods for the calculation of exponents of the lexical variety of texts" in: "Semiotika i iskusstvometrija", Moskva, "Mir", 1972). In this case reasonable results can only be achieved for small excerpts (that Vorončak used for a test of his model); the frequency structure of larger excerpts (when the low-frequency "tale" of the graph falls below the straight line, similar to the upper graph in fig. 4, see below), does not at all agree with this hypothesis.

are determined by means of an excerpt".

J. Carroll assumes that the word probabilities in the general lexical totality have a lognormal distribution (i.e. the logarithms of the word probabilities have a normal distribution). This point of view has been stated repeatedly in literature (for a review cf. Carroll, 1968) and it is indeed possible to approximate the frequency structures of large lexical excerpts by a lognormal curve with a cut-off. J. Carroll developed this assumption into a model in a number of publications of which Carroll (1969) is apparently the most comprehensive one; his model allows to calculate the increase in vocabulary and the frequency structures (analogous to V.M. Kalinin's constructions); besides in Carroll (1969) you will find a theoretical foundation of the lognormal hypothesis as well as results of a psycholinguistic experiment for its verification.

Unfortunately Carroll's calculating procedure is too complicated to be demonstrated here (in Carroll, 1969, part of it is presented as a computer program), but the figures he discusses look reliable. Kalinin's model and Carroll's model enable us to determine the most important parameter v - the dictionary size of the general lexical totality (the one whose word probabilities are distributed to the lognormal law) from the knowledge of some excerpt data. This is like a "potential dictionary" of a text, the number of different words with a non-zero probability of occurrence in a given text (provided of course that the hypothesis on the lognormal distribution is true). The weak point in Carroll's model is the bad description of the region of frequent words necessitating corrections that make the model more complicated. An explanation of how his model works will be given later on.

The structural model proposed by the author of this contribution and discussed in detail in Orlov 1976, 1978, was developed in the following way.

After it had become clear that dictionary size and frequency structure of complete texts of some important literary works are subject to the Zipf-Mandel'brot law (in the form 13 ÷ 15) the question arose of what happens in excerpts, why their dictionary is always larger than expected from (14) if the actual excerpt size is inserted for z.

If we assume that a text is a random sample taken from some unknown lexical totality the expected dictionary of the excerpt may be calculated by inserting the actually observed spectrum $v_m(N_o)$ in Kalinin's formula (1). The agreement was good (with an error of 3 ÷ 5%) so in this case Herdan's concept proved to be useful and the dictionary came up to the expectation that followed from Herdan's concept. But the question of the statistical structure of the text remained unsolved because formulas (1) and (2) require the knowledge of the actual observations for each excerpt size.

But why shouldn't the actually observed values of the frequency spectrum be replaced with the theoretical values calculated after (15), since they agree fairly well with the actual observations for a complete text? When inserting (14) and (15) into (1) we obtain:

(18) $$v(N, z) = v(z)\left(1 - \sum_{m=1}^{\infty} \frac{\left(1 - \frac{N}{z}\right)^m}{m(m+1)}\right)$$

The sum diverges if $\left|1 - \frac{N}{Z}\right| > 1$. But in this case we can find the analytical continuation; after a series of transformations the formula takes the form:

(19) $$v(N, z) = v(z)\frac{\ln \frac{z}{N}}{\frac{z}{N} - 1}$$

For the time being this means the following: it provides the dictionary size of a subexcerpt of size N taken from a text whose frequency structure for a size z exactly corresponds to the Zipf-Mandel'brot law (in the version 13 ÷ 15). Under the same conditions the analogous procedure of inserting (15) into (2) permits the calculation of the frequency spectrum of the subexcerpt; in this way one obtains rather lengthy expressions that may be written in a more compact form if the auxiliary paramter i_m is calculated - the number of different words each occurring at least m times in the subexcerpt of size N:

(20) $$i_m = \frac{v(z)}{\left(1 - \frac{z}{N}\right)^{m-1}} \left(\frac{\ln \frac{z}{N}}{\frac{z}{N} - 1} \quad \sum_{k=1}^{m-1} \frac{\left(1 - \frac{z}{N}\right)^{k-1}}{k} \right)$$

the frequency spectrum may then easily be determined from:

(21) $$v_m(N,z) = i_m - i_{m+1}$$

The accuracy of these formulas[1] with respect to excerpts of complete texts (whose size is inserted for z) is the same (± 20%) as that of the original formulas (13 ÷ 15) with respect to complete texts. It is self-evident that this is worse than calculations after (1) and (2) but on the other hand the application of formulas (19 ÷ 21) requires minimum information - only the knowledge of the length of the complete text and the frequency of the most frequent word therein (that may easily be estimated from a small excerpt). So there was no need to introduce parameters

---

1 In this paper it is impossible to talk about a number of complicated mathematical problems that arise in connection with inserting an idealized "Zipf's" frequency spectrum (15) in Kalinin's formulas (1) and (2). Let us only state that (15) may be regarded as a hypothetical expectation spectrum for size N. Thereby formulas (1) and (2) may now be used for predictions "forward" as well as "backward" (i.e. N may be greater as well as less than z in formulas (19 ÷ 21)).

into a model. The relative frequencies of frequent words were calculated after (13) as before as they are relatively stable, and the dictionary size of the text excerpt as well as the behavior of the "staircase" of the "tail" are calculated after (19) and (21).

To give an example we will analyze some data of table 2. The most interesting analysis is that of Dante's texts: Although the complete text of the "Divine Comedy" has a dictionary size that comes close to the dictionary size of a Zipf-text (after (14)) all three of its parts show considerable enlargements of the dictionary size (25%) compared with calculations after this formula. But if one calculates the dictionary size for the size of these parts and assumes that for the complete text of the "Divine Comedy" relations (13 ÷ 15) are applicable (i.e. z = 101,554 is inserted into (19), and for N the actual sizes of the parts are inserted), we obtain numbers in the range of 6,700 ÷ 6,750 which is in excellent agreement with reality (6,598; 6,450; 6,273) and all errors are less than 7%. If, however, the dictionary size of "La Vita Nuova" is calculated in the same manner, we obtain v (14,322; 101,554) = 4,100 (i.e. a similar dictionary size should be found in the excerpts of the "Divine Comedy" that correspond in size to "La Vita Nuova"). The actual dictionary size of this text is smaller by almost a factor of 2 (2,275) and differs from the theoretical dictionary size of a Zipf-text of the same length (2,225) only by 2%. That means that the lexical concentration of "La Vita Nuova" is strongly reduced in comparison with the "Divine Comedy" (and this again is only possible by increasing the fraction of frequent words in the text). It is this fact that leads to an exact validity of the Zipf-Mandel'brot law in each of the texts.

Formula (19) predicts the increase of vocabulary for "Pan Tadeusz" to the same degree of accuracy: for books I - VI the theoretical value of the dictionary size amounts to v (34,280; 64,510) = 6,050 which is calculated on the basis of the length of the complete text and differs from the actual dictionary size 6,828 by only 12%; as far as book I is concerned the theoretical value for v

(6,587; 64,510) = 2,200 differs from the actual dictionary size of 2,257 words by less than 3%. In this way the model actually describes the increase in vocabulary and the frequency structure of separate literary texts while at the same time there is no need to take any parameters into account a posteriori.

In its present form the model shows two fundamental disadvantages: a low degree of accuracy and a narrow range of application (only excerpts from literary works). Both disadvantages could be done away with by regarding z as a fitting parameter (i.e. the model would be changed into an approach with one parameter). If for an excerpt size N the dictionary $\hat{V}(N)$ and the frequency of the most frequent word are known, then z will be chosen in such a way that the following condition is fulfilled:

$$v(N,z) = \hat{v}(N) \qquad (22)$$

where the left-hand side v (N,z) is determined by formula (19). The solution of the transcendental equation (22) is described in detail in Orlov, 1978; cf. also the appendix. The obtained value for z will then be inserted into formulas (13, 14) and (17, 21) in which N is considered to be a continuous variable. From the geometrical point of view the calculations mean that we construct the curve of increase in dictionary size (19) starting from a single observed point (N, $\hat{v}(N)$) and from this construction we get the parameter z - the only one that has to be chosen. For such an approach the model requires practically the same amount of (original) information as the simplest descriptive models (with one parameter). In this case the value of the parameter z enables us to predict not only the increase in vocabulary (after formula (19)) but also the frequency structure (the range of words with high and medium frequencies after formula (13), the "staircase" of rare words after formulas (20, 21)) for text excerpts of arbitrary size N, as well "forward" and "backward" from the observed excerpt. An example of such prognoses is listed in table 3 (based on the material of Latvian newspaper texts that were already discussed, cf. table 1). The first variant of a prognosis was made from the data of an excerpt of ten thousand words (all prognoses in the "forward" direction); the prognosis for a text thirty times as long as the excerpt in "forward" direction the error amounts to nearly 10%. The second variant of a prognosis was made from the data of an excerpt of a hundred thousand words (analogous to the prognoses in table 1); the correspondence for this case is of course much better.

Table 3

N	$\hat{v}(N)$	$v(N, z)$	%	$v(N, z)$	%
10,000	2,688	2,701	+0.4	2,790	+3.8
20,000	4,249	4,102	−3.6	4,268	+0.4
40,000	6,193	5,995	−3.3	6,286	+1.5
60,000	7,629	7,352	−3.8	7,739	+1.4
100,000	9,834	9,316	−5.0	9,871*	+0.4
200,000	13,389	12,426	−8.1	13,268	−0.9
300,000	16,104	14,457	−10.2	15,500	−3.7
Initial data for prognosis		N 10,000 v 2,688 p_1 0.0347		N 100,000 v 9,834 p_2 0.0347	
Selected parameter values		z 50,300		z 57,760	

* The asterisk denotes the theoretical values of the dictionary size that have been made to coincide with actual values.

How are we to interpret the data or the statistical text-"construction" that emerge from an effective functioning of this model? The fact that one can establish the value of the parameter z for any (kind of) text allows the conclusion that for any text there is such an excerpt size z whose frequency structure is adequately described by formulas (13 ÷ 15), i.e. by the traditional form of the Zipf-Mandel'brot law. This size can of course be called "Zipf's size" as Zipf was the first to assume its existence and to try to determine its empirical value.[1] For all other sizes "deviations of the tails" are inevitable which are described by formulas (19 ÷ 21), that is to say the "canonical" form of the law proves to be temporary, only a special case of a larger spectrum of frequency structures (and only complete texts of separate literary works have a "tendency" towards this form "coming to a stop" near size z, for details cf. Orlov, 1978).

The parameter z also carries an immediate linguistic interpretation. The analysis of expression (19) shows that if we take two excerpts of the same length N from different contexts (for which different values of z apply) with the same value for p_1

(23) $$\upsilon(N_1 z_1) > \upsilon(N, z_2) \text{ if } z_1 > z_2$$

is valid.

---

1 It has to be conceded that Zipf himself apparently believed this size to be constant. For k in formula (9) he inserted 0.1 and concluded that if $\sum_{i=1}^{22{,}026} \frac{0.1}{i} = 1$, the size of the excerpt whose frequency structure is described by formula (9) should contain a dictionary with 22,026 words. If you bear in mind that the minimum absolute frequency of words equals 1, it is not difficult to find (s. below) that the dictionary of such an excerpt should contain approximately 220,000 words (it is revealing that these figures, according to Cheili's data, approach the size and dictionary of Joyce's novel "Ulysses" - Zipf's basic material). That is to say Zipf observed the fulfilment of "his" law just on the full length of a single work and thought this to be an inherent law of language in general; in (Nadarejšvili, Orlov, 1971; Orlov, 1978) you will find evidence for a difference in the frequency structure between "Ulysses" and other pieces of literature by Joyce. In contradistinction to this idea z can be of any value and its dictionary size is determined after formula (14) taking into account the constant B (and as well the deviation in the range of frequently used words on the log-log scale, cf. fig. 3). In particular, if we choose z = 220,000 and $P_1 = 0.1$ we get $V(z) = \frac{220{,}000}{\ln 22{,}000}$ 22,000, where $k = \frac{1}{\ln 22{,}000} = 0.1$ and $B = \frac{0.1}{0.1} - 1 = 0$, that means that Mandel'brot's "curve" (13) changes into Zipf's "straight line" (9). It would be more precise to call z "Zipf-Mandel'brot-quantity" but this would be very long.

This means that the observed dictionary of an excerpt is the larger the higher the value for z. As p_2 changes in comparatively narrow limits for texts in one language and as these changes have practically no effect on the applicability of the model into which p_2 enters logarithmically one can regard z as a relative *measure of lexical concentration*, of the relative lexical variety. The dynamics of alterations of this characteristic value for texts belonging to different epochs and genres (from some hundreds in Russian bylinas up to some hundreds of thousands in contemporary texts) is described in detail in Orlov (1978) on extensive material. There the function of the parameter z as an indicator of the statistical heterogeneity of texts is demonstrated.

If log v(N, z) is differentiated by log N, the derivative is

(24) $$L(x) = \frac{d\log V}{d\log N} = \frac{x}{x-1} - \frac{1}{\ln x}$$

with $x = \frac{Z}{N}$ giving the slope of the curve of increase in dictionary size plotted on a log-log scale. As can be seen from (24) $L(x) \rightarrow 0{,}5$ for $x \rightarrow 1$ (i.e. $N \approx z$). This is exactly the slope that Giro's ratio assumes for the curve of increase in dictionary size (cf. formula (6) on page 12). For relatively large deviations of N from z L(x) varies comparatively slowly. Therefore Giro's ratio may be accepted as a measure of lexical concentration (if N is of the same order as z for the text in question).

Orlov (1978) shows that the quantity L(x) also finds a linguistic interpretation - it is the differential rate of increase in dictionary size which means the ratio of the relative increase in dictionary size $\frac{\Delta V}{V}$ and the relative increment of the text size $\frac{\Delta N}{N}$ (N is the initial text size, N the increment of the text size, Δv the corresponding increase in dictionary size). At the same time L(x) is the ratio of words occurring once and the dictionary size. Therefore the validity of the Zipf-Mandel'brot law for complete texts of literary works means that the number of words occurring once represents about half of the stock of words (a frequent observation) and the text must develop more slowly roughly after the relative increase in dictionary size has fallen below one half of the increase in text size.

The disadvantage of this model is that it becomes unrealistic for very large N because the dictionary size increases indefinitely with increasing N. It is true, though, that practically the "very large N" are obviously far greater than the actual sizes of lexical excerpts. One may assume that within the model the formally observed irrelevant (on average for many texts and excerpts about 5 ÷ 8%, cf. (Orlov 1978)) enlargement of the number of words occurring once was caused by the fact that the model assumes an excerpt taken from a totality of infinite dictionary size, whereas the dictionary of real languages is finite. This is a further confirmation that the structural model (that the theory is mainly based upon) permits to gain positive statements even from its deviations from reality. We continue to discuss the investigated structural models taking the calculational error in data of the frequency dictionary (Kučera, Francis, 1967) as an example. The value of z is determined from the data of the dictionary of the whole body of texts by solving the transcendental equation (cf. (22))

$$V(1,014,232;z) = 50,306.$$

By the method of successive approximations (cf. appendix) $z \approx 280,000$ is found (i.e. especially for this value the Zipf-Mandel'brot law should apply most exactly: as among the "representative" excerpts of Francis and Kučera there is an excerpt of similar size of 253,533 words, this conclusion may directly be verified). Equally the frequency spectrum of the whole body of texts is used to calculate the increase in dictionary size after Kalinin's formula (1); this is the basic information for Carroll's model. The results of all theoretical calculations along with the actual data are presented in table 4.

As can be seen from the table the agreement is better among the

Table 4

Exerpt size	Actual dict. size	After Kalinin	After Carroll	After Orlov	After Tuldava	After Nesitoj
100	-	-	87	80	13	78
1,999	808	-	1,063	1,008	269	922
10,051	3,009	-	3,576	3,514	1,336	3,225
101,566	13,706	16,000	15,865	16,375	13,706*	13,700*
253,538	23,655	25,919	26,218	26,986	23,988	23,600*
400,000	-	32,614	-	33,136	32,258	30,650
1,014,232	50,406	50,406*	50,963*	50,435*	50,406*	51,000
10	-	-	118,624	104,377	75,529	158,000
10	-	-	206,309	167,263	79,529	420,000
	-	-	340,193		80,000	2.64·10
Initial data for prognosis		Actual whole spec-trum	Actual whole spec-trum	Excerpt size and actual dictionary size are listed in those lines that are marked by an asterisk.*		
Selected parameter values		No	μ =3.237 σ = 1.4136	z = =280,130	a = = 80,000 b = = 592,000	a_o = = 0.0147 k = = 0.00222

* The asterisk denotes the theoretical values of the dictionary size that have been made to coincide with actual values.

theoretical columns than between any theoretical and actual data. As Kalinin's formula (1) has a "model" character for random excerpts it can be maintained that the "representative" excerpts of Francis and Kučera are unjustly regarded as such. Their collections of texts from different categories in corresponding proportions as well as the whole body of material did not result in a sufficiently good "mixing", and the body of texts cannot be regarded as random subexcerpts from a corpus[1].

The good correspondence between two further models and the calculations after Kalinin's formula (1) makes evident that the ideas therein about the statistical text structure coincide well enough with reality. And this is so in spite of the fact that these ideas are completely different within the two models! One of the models (Carroll's) implies a finite "potential dictionary", the other model, however, an infinite one. But only for sizes of the order 10^8 a more serious difference between the prognoses of the two models becomes noticeable. This means that we can judge about the better adequacy of one of the models only if we have a hundred million (or more) excerpts available. It is by the way worthwhile mentioning that this is a rather artificial problem as the statistical variation of actual texts that manifests itself in the data discussed above may lead to an effacing of the results of excerpts of any size. When we choose a more homogeneous or a more "heterogeneous" body of texts we will get quite different values for the dictionary size.

This, of course, is a general weak point of structural models that

1 It is this fact which finds evidence in the alteration of the value z if it is calculated after Kučera's and Francis' data of the subexcerpts. If we calculate z after the data of the subexcerpt for N = 10,051 we get $z \approx 120{,}000$; for N = 101,566 z 150,000, for N = 253,538 $z \approx 200{,}000$. This increase of the value for z reflects an increase of the lexical concentration with the increase in size of the subexcerpts. Without this phenomenon the dictionary size of Kučera's and Francis' corpus would be considerably less; according to the data of the subexcerpt for N = 10,051 (the minimum "representative" subexcerpt) we get v (1,014,232; 12,000) = 32,300, i.e. less than the actual dictionary size by a factor of 1.5.

The variation of the parameter z with an increase of the text volume (which may also be achieved by combining similar excerpts) thus reflects the degree of its statistical heterogeneity (cf. also Orlov, 1978). Observations also show that a considerable increase of the number of words occurring once (by more than 10 ÷ 15%) can be achieved and this is compared to the calculations after formula (21) within excerpts that are deliberately from heterogeneous material.

have to rely on G. Herdan's concept. It is quite possible that a more profound investigation into the statistical characteristics of actual texts will after all permit to introduce "corrections concerning coherency" or "coefficients of diversity" into the structural models. For the time being the actual increase in dictionary size of coherent texts is more adequately described by formulas such as Ju.A. Tuldava's (7) or V.V. Nešitoj's (8) because of their taking into consideration additional information about the actual development of the curve of increase in dictionary size (the last two columns in table 4). Each of these curves was matched with *two* observed points of the actual increase in dictionary size (they are marked by an asterisk*; to put it differently: these models require a considerably larger amount of information before getting started - what is needed are data of at least two excerpts from one apparent totality). As follows from the table formula (7) gives a successful interpolation for the intermediate (between the initial points, with asterisk) point; the prognoses, however, as well "forward" as "backward" from the observed values are manifestly unsuccessful. If one were to accept the "forward" prognoses it would imply that the size of the body of texts according to Francis and Kučera nearly exhausted the dictionary size of the lexical totality. Formula (8) works much better but this time the "forward" prognoses are a little too high.

Fig. 4 demonstrates a possibility that only structural models are capable of - the prognosis of the frequency structure (based on the data of Francis' and Kučera's dictionary). The upper curve shows the structure of the frequency curve and the "staircases" of the frequency spectrum at both ends: at the frequency range of the most frequent word p_1 and at the actual dictionary size (the right end of the lower step). The part with a continuous slope is calculated after formula (13) for z = 280,000 and P_1 = 0.069 (the obtained values were normalized for the size N = 1,014,232). The last parts of the steps on the lower right side that are drawn in thin lines were directly calculated after formula (20); the fat "landings" represent actually observed frequencies of rare words.

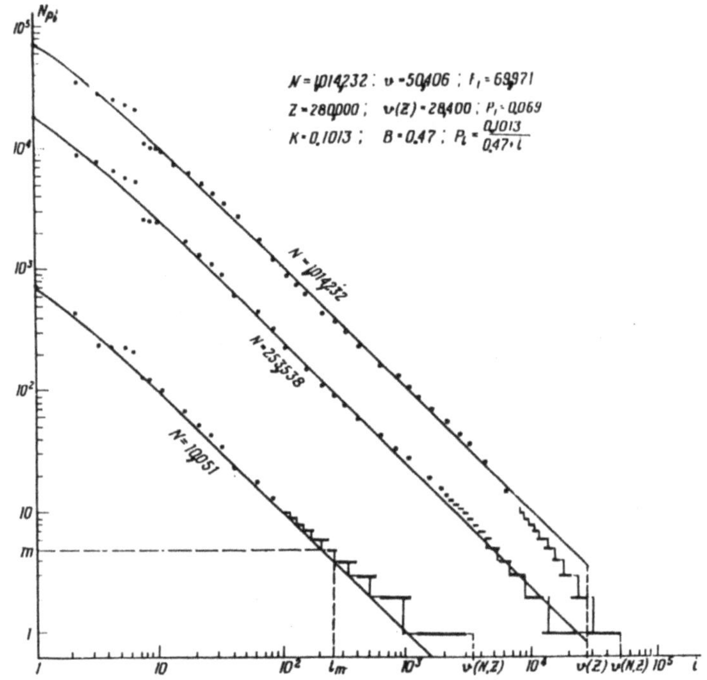

Fig. 4

The medium curve gives a prognosis for the size N = 253,538 for the same values z and p_1 that were determined on the basis of the whole body of texts; the calculations after formula (13) were now adapted to fit the size of this subexcerpt. For this case the "staircase" of rare words is (as well theoretically as empirically) much more rectilinear and up to the end coincides with curve (13), i.e. the conclusion is confirmed that for sizes near z = 280,000 the Zipf-Mandel'brot law is relatively well fulfilled. The lower curve represents an analogous prognosis for N = 10,051; here the "tail" of the graph of words occurring with low frequencies is clearly deflected upwards - in complete correspondence with the theoretical prognosis.

Thus the frequency structure of the excerpts is dynamic and depends on the size of the excerpt; and all calculations of the exponent γ in Mandel'brot's formula (11) carried out separately for different zones of the frequency graph are unsuccessful since they do not characterize anything but the given excerpt and the author's taste (into which zones the graph is to be divided). The proposed model does not only cope with the traditional difficulties of describing such graphs but is also capable of a conclusive prognosis of the frequency structure for excerpts differing in size by a factor of one hundred (in the present case). Roughly the same prognosis is possible with Carroll's model (cf. graph $E_{V''}$ on p. 424 in Kucera, Francis, 1967) though the correspondence is not so good for small sizes and in the range of higher frequencies (though it is difficult to make comparisons as Carroll's graphs are plotted in a different coordinate system). It is as well necessary to underline that Carroll's model requires full information about the frequency structure of the "basis"-size while the model (13 $\div$ 15, 19 $\div$ 21) "manages" with data on the dictionary size and the frequency of the most frequent word (for rough prognoses you may use the average value of the largest word probability p_1 = 0.05).

General recommendations that result from the assessment of the models discussed so far can be summarized in the following way:

If a reliable prognosis with reasonable accuracy is to be achieved, a complete description of the increase in dictionary size and the frequency structure of a text, a clear mathematical or linguistic interpretation of the parameters (or of the deviations between theory and experiment) - then structural models should be made use of; but if a most exact description of the observed data is desirable the descriptive models with many parameters become interesting but a prognosis (i.e. everything that is beyond observed facts) for this case is not reliable and the interpretation of the parameters becomes more difficult (an interpretation of the deviations between theory and experiment is not at all possible). Descriptive models with one parameter are suitable for rough "estimates", as long as their range of applicability is not left. In case a most reliable means is needed for examing models of any kind it is advisable to use relationships (1 ÷ 3) that permit to transform models for the frequency spectrum into models for the increase in dictionary size and vice versa and as well to control (and correct) the influence of the statistical non-random character of actual coherent texts.

The author would like to thank Professor R.G. Piotrovskij who inspired this work that deals with the comparative analysis of different models on the statistical text organization, he is also grateful to Dr. Ju.A. Tuldava and Dr. V.V. Nešitoj for close co-operation in understanding the functioning and possibilities of their models.

Appendix

Numerous discussions of the proposed model on the lexical frequency structure convinced the author that the "stumbling stone" of its application became the solution of equation (22) that supplies the value of the parameter z for the data of a concrete text. The iteration of this equation is described in Orlov (1978), where nomograms are also presented to get the first approximation, nomograms, that allow to quickly achieve a fairly exact solution (as a rule not more than three iterations). Unfortunately these nomograms do not supply all possible combinations of the values for N, v and z and without a "good" first approximation the number of iterations mounts up to 6 - 8. That is why in this paper we are going to describe the more usual process that does not require a good first approximation. If the size of the excerpt N is known, the size of its dictionary v and the frequency of the most frequent word p_1, then the calculation of z follows Diagram 1. At first the quantity k_o may assume any value but it is reasonable to choose as "zero-order approximation" its "typical" value, let us say $k_o = 0.1$.

As an example we calculate z for the data of Kučera's and Francis' dictionary.

The following quantities are given: N = 1,014,232; v = 50,406; $p_1 = 0.069$; $k_o = 0.1$.

1. D = 1,014,232/50,406 = 20.1213.

2. V = 20.1213 · 0.1 = 2.0121.

3. From table 5 we get the auxiliary quantities a and b. As step 2 yields values between 1.4138 and 2.1140 for Y we take from table 5: a = 0.0676 and b = 1.7164; now X can be calculated:

$$X = e^{-(0.0676 + 1.7164 \ln 2.0121} = e^{-1.2676} = 0.2815$$

4. z = 1,014,232 · 0.2815 = 285,506

5. $k = \frac{1}{\ln (285,506 \cdot 0.069)} = \frac{1}{\ln 19,700} = 0.1011.$

6. When we compare the obtained value k with the initial value $k_o = 0.1$ we see that it is increased by 0.0011; therefore we continue with step 7.

7. We set $k_o = 0.1011$.

2'. Y = 20.1213 · 0.1011 = 2.0344

3'. The new value Y belongs to the same interval in table 5 as the previous one; so we use the same figures a and b:

$$X = e^{-(0.0676 + 1.7164 \ln 2.0344)} = e^{-2.2865} = 0.2762.$$

4'. z = 280,131

5'. $k = \frac{1}{\ln 19,329} = 0.1013$

6'. This value differs from the previous one by only 0.0002, therefore we terminate the iteration. Further calculations (steps 8 - 10) serve as a final check of the correctness of all calculations; at the same time the constants B and v(z) are calculated:

8. B = 0.1013/0.069 - 1 = 0.4681.

9. $v(z) = 0.1013 \cdot 280{,}131 = 28{,}377.$

10. $v(N, z) = 28{,}377 \dfrac{\ln 0.2762}{0.2762 - 1} = 50{,}435$

Table 5

Y	*a*	*b*	*Y*	*a*	*b*
0,1330			0,9950		
	7,1553	7,2710		0,0018	1,8977
0,1454			1,4138		
	5,8302	6,5837		0,0676	1,7164
0,1612			2,1140		
	4,6863	5,9570		0,1747	1,5733
0,1811			3,2983		
	3,6018	5,3223		0,3019	1,4642
0,2053			5,2709		
	2,6315	4,7095		0,4427	1,3813
0,2390			8,7016		
	1,8148	4,1389		0,5803	1,3177
0,2814			14,720		
	1,1633	3,6250		0,7118	1,2688
0,3107			25,536		
	0,6566	3,1514		0,8346	1,2309
0,4248			44,822		
	0,3033	2,7418		0,9161	1,2015
0,5464			79,597		
	0,0946	2,3966		1,0488	1,1781
0,7191			110,12		
	0,0060	2,1276		1,1386	1,1590
0,9950			270,00		

The obtained theoretical value for the dictionary size for an initial size N = 1,014,232 differs from the actual one (50,406) by only 0.06%; this means that all calculations have been carried out correctly. The general accuracy (which is limited by the accuracy of table 5) of the described method is such that in adverse cases the relative deviation between the right hand side and the left hand side of equation (22) may amount to approximately 1%; in the case of larger deviations the error should lie in the calculation.

It has to be underlined that "adverse cases" occur when z is calculated from the data of small excerpts whose dictionary consists of only a few hundred words. In this case the absolute deviation between the right hand side and the left hand side of equation (22) (in the calculation of z by the described method) is less than 10 words taking into account that the fluctuation of the dictionary size in excerpts of this size is considerably larger this degree of accuracy has to be regarded as quite satisfactory.

If it is desirable for some reason to increase the accuracy in the calculation of z this can be achieved by applying the trapezoidal rule because its convergence is not limited.

Once two "good" approximations z_{i-1} and z_i are known all further approximations may be found according to

$$(25) \qquad z_{i+1} = z_{i-1} + \frac{v - v(N, z_{i-1})}{v(N, z_i) - v(N, z_{i-1})} = (z_i - z_{i-1}).$$

where v stands for the actual dictionary size for size N as before.

The trapezoidal rule can also be recommended for a machine realization (in this case its slow convergence does not mean an important disadvantage). The first approximation may then be a universal constant (for example, $z_1 = 10^5$), the second approximation may be taken to be less or greater by a factor of two depending on the sign of the deviation $\Delta_1 = v - v(N, z_1)$ ($z_2 = 5 \cdot 10^4$ for $\Delta_1 < 0$; $z_2 = 2 \cdot 10^5$ for $\Delta_1 > 0$), the third and all following are determined by (25). This procedure may be continued until a certain value $|\Delta_i|$ is reached or the value $\frac{|\Delta_1|}{v}$ depending on the desired accuracy.

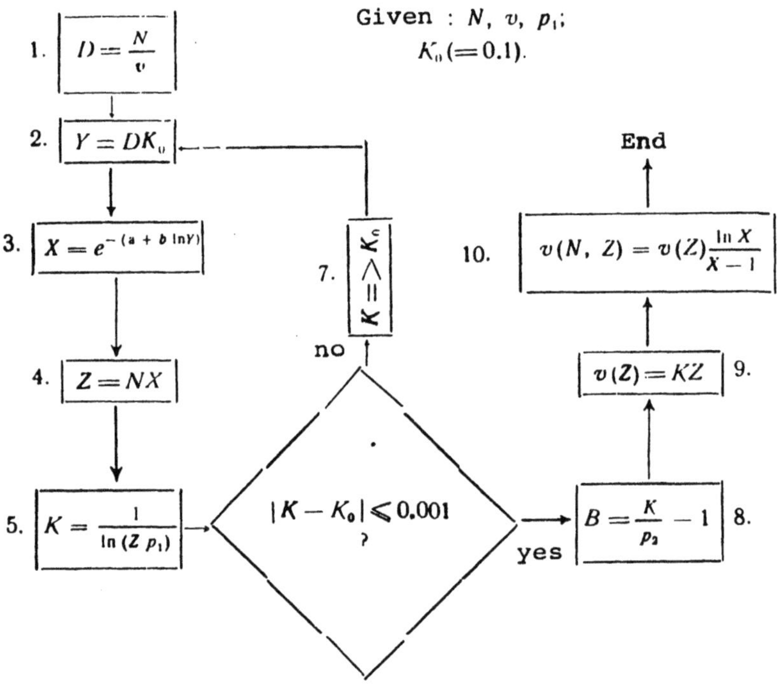
Given : N, v, p_1;
$K_0 (= 0.1)$.
1. $D = \frac{N}{v}$
2. $Y = DK_0$
3. $X = e^{-(a + b \ln Y)}$
4. $Z = NX$
5. $K = \frac{1}{\ln (Z p_1)}$
$|K - K_0| \leqslant 0.001$?
no
7. $K => K_0$
yes
8. $B = \frac{K}{p_2} - 1$
9. $v(Z) = KZ$
10. $v(N, Z) = v(Z) \frac{\ln X}{X - 1}$
End

Diagram 1

Bibliography (Russian references)

Abdulaeva, P.A., Bektaev, K.B., Beljaeva, L.N., Kobzeva, A.T., Krisevič, V.S., Moskvina, V.N., Prichod'ko, L.P., Sadčikova, P.V., Šamšatova, G.I., Čajkovskaja, I.I., *Opyt realizacii promyšlennogo MP tekstov na russkij jazyk.*-APTMPL, 1977.

Alekseev, P.M., *Leksičeskaja i morfologičeskaja statistika anglijskogo pod"jazyka elektroniki.*-StR, 1968.

Alekseev, P.M., *Statističeskaja leksikografija (tipologija, sostavlenie i primenenie častotnych slovarej)*.L., 1975.

Alekseev, P.M., Byčkov, V.N., *O statističeskich modeljach bol'šich tekstovych massivov dlja nužd mašinnogo perevoda i informacionnogo poiska.* APTMPL, 1977.

Alekseev, P.M., Kaširina, M.E., Tarasova, E.M., *Častotnyj anglo-russkij fizičeskij slovar'-minimum.* M., 1979.

Arapov, M.V., Efimova, E.N., Šrejder, Ju.A., *O smysle rangovych raspredelenij.*-NTI, 1975, No. 1.

Arapov, M.V., Efimova, E.N., Šrejder, Ju.A., *Rangovye raspredelenija v tekste i jazyke.*-NTI, 1975, No. 2.

Arapov, M.V., Efimova, E.N., *Ponjatie leksičeskoj struktury teksta.* NTI, serija 2, 1975, No. 6.

Arapov, M.V., Cherc, M.M., *Matematičeskie metody v istoričeskoj lingvistike*.M., 1974.

Arzikulov, Ch.A., Popesku, A.N., Rachubo, N.P., Saitkulov, Ch.S., Chažinskaja, M.S., Čižakovskij, V.A., *Tezaurusnoe raspoznavanie smysla francuzskich naučno-techničeskich tekstov.*-Lingvističeskie issledovanija-1976. Voprosy leksikologii, leksikografii i prikladnoj lingvistiki. M., 1976.

Barabaš, Ju.L., Varskij, B.V., Zinov'ev, V.T., Kiričenko, V.S., Sanegin, V.F., *Voprosy statističeskoj teorii raspoznavanija.* M., 1967.

Berztiss, A.T., *Struktura dannych.* M., 1974.

Beljaeva, L.N. i dr., *Mašinnaja morfologija v MARS.*-Lingvističeskie issledovanija-1976.M., 1976.

Bektaev, K.B., *Statistiko-informacionnaja tipologija tjurkskogo teksta*.Alma-Ata, 1978.

Bektaev, K.B., Belocerkovskaja, L.I., Markus, E.G., Piotrovskij, R.G., Sadčikova, P.V., Uchalkina, A.M., *Promyšlennyj mašinnyj perevod anglijskich, francuzskich patentnych tekstov po nizkotemperaturnoj plazme na russkij jazyk. Materialy k VII Vsesojuznoj konferencii po generatoram nizkotemperaturnoj plazmy.* Tom II. Alma-Ata, 1977.

Bektaev, K.B., Luk'janenkov, K.F., *O zakonach raspredelenija edinic pis'mennoj reči.*-StRAAT, 1971.

Bektaev, K.B., Piotrovskij, R.G., *Matematičeskie metody v jazykoznanii. C.I-II.* Alma-Ata, 1973-1974.

Belonogov, G.G., Bogatyrev, V.I., *Avtomatizirovannye informacionnye sistemy.* M., 1973.

Berzon, V.E., *Nekotorye sposoby formalizacii processa ustanovlenija otnošenija i zavisimosti meždu predloženijami v svjaznom tekste.* -NTI, 1971, No. 8.

Bilan, V.N., Krisevič, V.S., Nizamutdinova, I.G., Piotrovskij, R.G., *Semantiko-sintaksičeskij perevod v gruppe "Statistika reči".* - AAT, 1976.

Blekuell, D., Giršik, M.A., *Teorija igr i statističeskich rešenij.* Russk. per. M., 1956.

Bol'šev, L.P., Smirnov, N.V., *Tablicy matematičeskoj statistiki.* M., 1965.

Borisevič, A.D., Gončarenko, V.V., Grečiškina, V.I., Dobruskina, Ė.M., i dr., *Kodirovanie grammatičeskoj informacii v mašinnom slovare.* - Statistika teksta, t.II. Avtomatičeskaja pererabotka teksta. Minsk, 1970.

Boroda, M.G., Nadarejšvili, I.S., Orlov, Ju.K. Čitašvili, R.Ja., *O charaktere raspredelenija informacionnych edinic maloj častoty v chudožestvennych tekstach.*-Semiotika i informatika, vyp. 9, 1977.

Brudnyj, A., Šrejder, Ju., *Kommunikacija i intellekt.*-Meždunarodnyj forum po informacii i dokumentacii, tom I, No. 2, M., 1976.

Bulachov, M.G., *Materialy dlja častotnogo slovarja russkogo jazyka.* "Leksikalogija i gramatyka". Minsk, 1969.

Vasil'ev, V.I., *Raspoznajuščie sistemy.* Kiev, 1969.

Vertel', V.A., Vertel', E.V., Krisevič, V.S., Piotrovskij, R.G., Tribis, L.I., *Avtomatičeskie slovari v sisteme binarnogo verojatnostnogo MP.* -Inženernaja lingvistika. Učenye zapiski LGPI im.A.I. Gercena, t. 438, c. I-II, L., 1971.

Vertel', E.V., Šingareva, E.A., *IPS "Otvet".* APTMPL, 1977.

Vinograd, T., *Programma, ponimajuščaja estestvennyj jazyk.* Russk. per. M., 1976.

Volochov, V.S., Zajčenko, Ju.P., *Dispersionnyj metod samoproizvol'nogo (spontannogo) razbienija prostranstva izobraženij na kompaktnye podmnožestva (obrazy).*-Avtomatika, 1966, No.5; 1969, No. 1.

Volochov, V.S., Popesku, A.N., *Metod avtomatičeskogo razbienija semantičeskogo prostranstva na kompaktnye oblasti.* -Metody analiza teksta. Minsk, 1975.

Gil'denberg, E.I., *K statisticeskoj charakteristike leksiki "Geroja našego vremeni".* "Typalogija i uzaemadzejanne slovjanskich mou i litaratur". Minsk, 1973.

Gindin, S.I., *Ontologičeskoe edinstvo teksta i vidy vnutrifrazovoj organizacii.*-Mašinnyj perevod i prikladnaja lingvistika, vyp. 14, M., 1971.

Gnedenko, B.V., *Kurs teorii verojatnostej.* M., 1965.

Gorelik, A.L., Skrupkin, V.A., *Nekotorye voprosy postroenija sistemy raspoznavanija.* M., 1974.

Grenander, U., *Slučajnye processy i statističeskie vyvody.* Russk. per. M., 1961.

Grin, K., *Dokazatel'stvo teorem s ispol'zovaniem pravil rezoljucii kak osnova dlja postroenija voprosno-otvetnoj sistemy.* Russk. per.v kn.: Slėjgl, Dz., *Iskusstvennyj intellekt. Podchod na osnove ėvrističeskogo programmirovanija.* Russk. per. M., 1973.

Zade, L.A., *Osnovy novogo podchoda k analizu složnych sistem i processov prinjatija rešenij.*-Matematika segodnja. No. 7, serija "Matematika. Kibernetika", M., 1974.

Zade, L., *Ponjatie lingvističeskoj peremennoj i ego primenenie k prinjatiju približennych rešenij.* Russk. per. M., 1976.

Zaikin, O.A., Luk'janova, E.M., *Principy postroenija bazy dannych dlja avtomatizirovannych sistem naučno-techničeskoj informacii.*- Izvestija Leningradskogo ordena Lenina Ėlektrotechničeskogo instituta im. V.I. Ul'janova (Lenina). Vyp. 214, L., 1977.

Zubov, A.V., *Obrabotka na ES ĖVM tekstov estestvennych jazykov.* Minsk, 1977.

Zubov, A.V., Luk'janenkov, K.F., Piotrovskij, R.G., Chotjašov, E.P., *Leksiko-statističeskoe opisanie teksta na elektronno-vyčislitel'nych mašinach.*-StR, 1968.

Zueva, T.R., Sova, L.Z., *Referencionnaja klassifikacija russkich suščestvitel'nych*.-Lingvističeskie issledovanija. Voprosy stroja indoevropejskich jazykov. Č. 2. M., 1975.

Ivankin, V.I., *Semantiko-distributivnyj metod sostavlenija spiskov ključevych slov, rekomenduemych dlja vvoda v poiskovye obrazcy dokumentov*.-NTI, 1974, No. 10. Informacionnye sistemy obščego naznačenija. M., 1975.

Kabalkin, E.I., Michajlova, I.V., Piotrovskij, R.G., Rožkov, V.V., Sodomov, Ju.B., *Pererabotka estestvennogo jazyka konečnym avtomatom v uslovijach adaptivnogo dialoga*.-VSVA, 1976.

Kabalkin, E.I., Piotrovskij, R.G., Chochlov, L.M., *Prikladnye aspekty obucenija avtomatov estestvennym jazykam*.-VSVA, 1976(a).

Kabalkin, E.I., Piotrovskij, R.G., Chochlov, L.M., *Prikladnye aspekty obučenija avtomatov estestvennym jazykam*.-Inženernaja lingvistika i optimizacija prepodavanija inostrannych jazykov. L., 1976(b).

Kalinin, V.M., *Nekotorye statističeskie zakony matematičeskoj lingvistiki*.-Problemy kibernetiki, vyp. II, M., 1964.

Kalinin, V.M., *Funkcionaly, svjazannye s raspredeleniem Puassona i statističeskaja struktura teksta*.-Trudy matematič. in-ta im. Steklova, LXXIX, M.-L., 1965.

Kaširina, M.E., *O tipach raspredelenija leksičeskich edinic v tekste*.-StRAAT, 1974.

Kvaracchelija, G.Š., *Opyt sravnitel'no-stilističeskogo analiza leksiki statističeskimi metodami*.-Voprosy sovremennogo obščego i matematičeskogo jazykoznanija, II, 1967, (na gruz. jaz.).

Kolguškin, A.N., *Lingvistika v voennom dele*. M., 1970.

Korn, G., Korn, T., *Spravočnik po matematike*. M., 1970.

Krutikova, O.F., *Nekotorye statističeskie ocenki nizkočastotnych slov*.-Issledovanija po informacii, M., 1968.

Kuzin, L.T., *O sistemach iskusstvennogo intellekta*.-IIIP, 1974.

Leonov, V.P., *O metodach avtomatičeskogo referirovanija*.-NTI, 1976, No. 6.

Luk'janova, E.M., *Strukturirovanie bazy dannych na osnove poiskovoj matricy lingvističeskogo algoritma*. Lingvističeskie issledovanija- 1977, M., 1978.

Luk'janova, E.M. i dr., *Razrabotka struktur dannych s ispol'zovaniem apparata k-formul*.-Trudy NIIUMS, Perm', vyp. XVIII, 1978.

Martin, Dž., *Organizacija bez dannych v vyčislitel'nych sistemach*. M., 1978.

Mandel'brot, B., *Teorija informacii i psicholingvistika: teorija častot slov*.-Matematičeskie metody v social'nych naukach. Russk. per. M., 1973.

Mandel'brot, B., *O rekurrentnom kodirovanii, ograničivajuščem vlijanie pomech*. Russk. per. Teorija peredači soobščenij. M., 1957.

Markus, S., *Teoretiko-množestvennye modeli jazykov*. M., 1970. Matematičeskie metody v jazykoznanii. Riga, 1969.

Midou, Č., *Analiz informacionnych sistem*.Russk. per. M., 1977.

Mikerina, T.A., *Nekotorye statističeskie priemy leksiko-morfologičeskogo opisanija funkcional'nogo stilja (na materiale anglijskich tekstov po sudostroeniju)*. AKD. L., 1967.

Nadarejšvili, I.Š., Orlov, Ju.K., *Rost leksiki kak funkcija dliny teksta (na primere proizvedenij L.N. Tolstogo i D. Džojsa)*.-Soobščenija AN Gruz. SSR. Tbilisi, 1971, t. 64, No. 3.

Nadarejšvili, I.Š., Orlov, Ju.K., *Metod polnoj fiksacii teksta pri lingvostatističeskom analize*.-Linguistica X, Tartu, 1977.

Nadarejšvili, G.S., Nadarejšvili, I.S., Orlov, Ju.K., *Nestacionarnye javlenija v processe poroždenija teksta*.-Trudy Instituta kibernetiki AN GSSR, t. I, Tbilisi, 1977.

Neljubin, L.L., *Informacionno-statističeskie i lingvističeskie osobennosti jazyka i teksta v uslovijach speckommunikacii*. ADD. M., 1975.

Nemirovskij, E.L., *Opyt postroenija binarnoj klassifikacii*.-NTI, 1971, No. 1.

Nešitoj, V.V., *Zakony raspredelenija slov v tekste i ego leksičeskaja parametrizacija*. AKD. Minsk, 1973.

Nešitoj, V.V., *Ocenka leksičeskoj blizosti tekstov*. V kn.: Deduškin, A.N., Nešitoj, V.V., Cvemetidze, S.V., Razrabotka avtomatizirovannych sistem naučno-techničeskoj informacii. Minsk, 1976.

Nešitoj, V.V., *Raspredelenie ključevych slov v tekste*. "Kibernetika", 1977, No. 2.

Orlov, Ju.K., *Obobščenie zakona Zipfa*.-Problemy prikladnoj lingvistiki. Tezisy dokladov mežvuzovskoj konferencii, Moskva, 1969.

Orlov, Ju.K., *Obobščenie zakona Zipfa-Mandel'brota*-Soobščenija AN GSSR, t. 57, No. 1, 1970.

Orlov, Ju.K., *O statističeskoj strukture soobščenij, optimal'nych dlja čelovečeskogo vosprijatija.*-NTI, 1970, No. 8.

Orlov, Ju.K., *Častotnye struktury konečnych soobščenij v nekotorych estestvennych informacionnych sistemach.* AKD. Tbilisi, 1974.

Orlov, Ju.K., *Obobščennyj zakon Zipfa-Mandel'brota i častotnye struktury informacionnych edinic različnych urovnej.* -Vyčislitel'naja lingvistika. M., 1976.

Orlov, Ju.K., *Model' častotnoj struktury leksiki.*-Issledovanija v oblasti vyčislitel'noj lingvistiki, vyp. II, č. I, M., 1978.

Piotrovskaja, A.A., Piotrovskij, R.G., *Iskusstvennyj intellekt i jazykoznanie.* Zeitschrift für Phonetik, Sprachwissenschaft und Kommunikationsforschung. B. 31, H. 1, 1978.

Piotrovskij, R.G., *Tekst, mašina, čelovek.* L., 1975.

Piotrovskij, R.G., *Avtomatičeskaja pererabotka teksta: teorija, eksperiment, vnedrenie.*-APTMPL, 1977.

Piotrovskij, R.G., *Inženernaja lingvistika i iskusstvennyj intellekt.*-Modelirovanie informacionnych processov celenapravlennogo povedenija. (Tezisy simpoziuma. Tbilisi, 9-12 nojabrja, 1976 g.) Tbilisi, 1976.

Piotrovskij, R.G., *Inženernaja lingvistika: teorija, ėksperiment, realizacija.* Izvestija AN SSSR, serija literatury i jazyka. T. 37, No. 1, 1978.

Piotrovskij, R.G., Bektaev, K.B., *Mašinnyj perevod: teorija, ėksperiment, vnedrenie.* VJa, 1977, No. 5.

Piotrovskij, R.G., Bektaev, K.B., Piotrovskaja, A.A., *Matematičeskaja lingvistika.* M., 1977.

Popesku, A.N., *Avtomatičeskoe indeksirovanie i annotirovanie naučnotechničeskich tekstov.*-Vsesojuznyj seminar po informacionnym jazykam. Predvaritel'nye publikacii. Vyp. 4, M., 1972.

Popesku, A.N., Chažinskaja, M.S., *Tezaurusnyj metod avtomatičeskogo raspoznavanija smyslovogo obraza naučno-techničeskich tekstov.*-Lingvostatistika i avtomatičeskij analiz tekstov. Minsk, 1972.

Pospelov, G.S., Pospelov, D.A., *Iskusstvennyj intellekt: sostojanie i problemy.*-IIIP, 1974.

Programma po rešeniju naučno-techniceskoj problemy 0.80.018 (SĖV) "Sozdat' gosudarstvennuju avtomatizirovannuju sistemu naučno-techničeskoj informacii, obespečivajuščuju povyšenie naučno-techničeskogo urovnja i ėffektivnosti issledovanij, razrabotok i proizvodstva na 1976-1980 gg."

Rejtman, U., *Poznanie i myšlenie. Modelirovanie na urovne informacionnych processov.* Russk. per. M., 1968.

Sadčikova, P.V., *Anglijskij i russkij častotnye slovari po chimii polimerov.*-StRAAT, 1974.

Sistema matematičeskogo obespečenija "Bank dannych universal'noj struktury". Kalinin, Centrprogrammsistem, 1975.

Seger, N., *Sintaksičeskij analiz estestvennogo jazyka.*-Avtomatičeskij perevod. M., 1971.

Sinenko, G.D., *Častotnyj slovar' "Povesti o detstve" F.V. Gladkova i voprosy literaturnogo stilja.*-Vesnik Belaruskaga dzjaržaunaga universiteta imja U.I. Lenina. Seryja IV, No. 1, 1973.

Skorochod'ko, Ė.F., *Semantičeskie svjazi v leksike i tekstach.*-VITP, sb. M., 1974(a).

Skorochod'ko, E.F., *Lingvističeskie problemy obrabotki tekstov v avtomatizirovannych informacionno-poiskovych sistemach.*-VITP, sb. No. 25. M., 1974(b).

Sova, L.Z., *Raspoznavanie smyslovych svjazej pri avtomatičeskoj obrabotke teksta*-Metody analiza tekstov. Minsk, 1975.

Sova, L.Z., *Avtomatičeskoe raspoznavanie smyslovogo obrazca v sintaksise.*-StRAAT, 1974.

Stepanov, Ju.S., *Osnovy obščego jazykoznanija.* M., 1975.

Sudavičene, L., *Iz opyta sostavlenija častotnogo slovarja jazyka K. Paustovskogo.*-Uč. zapiski vysšich uc. zavedenij Lit. SSR Jazykoznanie. XXII (2). Vil'njus, 1971.

Terešenkova, A.V., Bagrinovskaja, A.K., *K voprosu postroenija logičeskich struktur bez dannych.* Algoritmy i organizacija rešenija ėkonomičeskich zadač, vyp. II, M., 1978.

Tokarev, V.P., Jakubajtis, T.A., *Matematiko-statističeskaja model' raspredelenija leksem.*-Matematičeskie metody v jazykoznanii. Riga, 1969.

Tuldava, Ju.A., *O statističeskoj strukture teksta.*-Sovetskaja pedagogika i škola. XI, Tartu, 1974.

Turygina, L.A., Borkun, M.A., *Statističeskie metody issledovanija častotnogo raspredelenija lingvističeskich edinic.* Ėntropija jazyka i statistika reči. Minsk, 1966.

Ugodčikova, N.F., Kuznecova, G.E., Zubov, A.V., *Algoritm i programma razrešenija reguljarnoj leksičeskoj mnogoznačnosti suščestvitel'nych*.-AAT, 1976.

Uilks, U., *Matematičeskaja statistika*. Russk. per. M., 1967.

Uspenskij, V.A., *Odna model'dlja ponjatija fonemy*.-Voprosy jazykoznanija, 1974, No. 6.

Fel'dbaum, A.A., Dudykin, A.L., Manovcev, A.P., Miroljubov, N.P., *Teoretičeskie osnovy svjazi i upravlenija*. M., 1963.

Fu, K., *Posledovatel'nye metody v raspoznavanii obrazov i obučenija mašin*. Russk. per. M., 1971.

Chažinskaja, M.S., *Tezaurusnaja semantičeskaja set' v dialogovych AIPS*.-APTMPL, 1977.

Chellstrom, K., *Statističeskaja teorija obnaruženija signalov*. Russk. per. M., 1963.

Calenko, M.Š., *Reljacionnye modeli bez dannych (obzor)*.-Algoritmy i organizacija rešenija ėkonomičeskich zadač. Vyp. 9, M., 1977.

Častotnyj slovar' russkogo jazyka. M., 1977.

Šenk, R., Abel'son, R., *Scenarii, plany i znanija*.-Trudy IV Meždunarodnoj ob"edinennoj konferencii po iskusstvennomu intellektu. M., 1975.

Šingareva, E.A., *Opyt ispol'zovanija frejmov dlja raspoznavanija smyslovogo obraza teksta*.-NTI, 1978, No. 10.

Šrejder, Ju.A., *Ravenstvo, schodstvo, porjadok*. M., 1971.

Šrejder, Ju.A., *Logika klassifikacii*.-NTI, 1973, No. 5.

Šrejder, Ju.A., *Logika znakovych sistem*.-Matematika segodnja, serija "Matematika, Kibernetika". M., 1974, No. 1.

Šrejder, Ju.A., *O vozmožnosti teoretičeskogo obosnovanija statističeskich zakonomernostej teksta*.-Problemy teorii peredači soobščenij. M., 1966.

Non-russian references

A Concordance to Byron's Don Juan. Ithaca, 1967.

Allen, S., *Nusvensk frekvensordbok baserad på tidningstext*. Stockholm, 1970-1971.

Alinei, M., *Spogli elettronici de l'Italiano delle origini e del duecento*. Il Eorma 5, 8, 1971.

Bazin, U., *Index du vocabulaire de "l'Etranger" d'A.Camus.* NIZET, Sorbonne, Paris, 1969.

Berul, L., *Document Retrieval.*-AR, vol. 4, 1969.

Bookchin, B., *Computer Outputs for Sentence Decomposition of Scientific Texts.* New York University, Institute for Computer Research in the Humanities, Linguistic String Project, New York, March 1968. (String program reports, no 3).

Booth, A.D., *A "Law" of Occurrences for Words of Low Frequency.*- Information and Control, v. 10, N 4, 1967.

Carroll, J.B., *On Sampling from a Lognormal Model of Word-Frequency Distribution.* In: Kučera, H., Francis, W.N., *Computational Analysis of Present-Day American English.* Providence, 1967.

Carroll, J.B., *Statistical Analysis of the Corpus.* In: Carroll, J.B., Davies, P., Richman, B., *The American Heritage Word Frequency Book.* New York, 1971.

Carroll, J.B., *A Rationale for an Asymptotic Lognormal Form of Word-Frequency Distributions.*-Research Bulletin. Educational Testing Service. Princeton, Nov. 1969.

Carroll, J.B., *Word-Frequency Studies and the Lognormal Distribution.*-Proceedings of the Conference on Language and Language Behavior (ed. by E.M. Zale). New York, 1968.

Carroll, J.B., Davies, P., Richman, B., *The American Heritage Word Frequency Book. New York, 1971.*

Dictionnaire des fréquences. Vocabulaire littéraire des XIX et XX siècles. Paris, 1971.

Fairthorne, R., *Content Analysis, Specification and Control.*- AR, vol. 4, 1968.

Guiraud, P., *Les caractères statistiques du vocabulaire. Essai de méthodologie.* Paris, 1954.

Herdan, G., *Type-Token Mathematics.* 's-Gravenhage, 1960.

Herdan, G., *The Advanced Theory of Language as Choice and Chance.* New York, 1966.

Kučera, H., Francis, W.N., *Computational Analysis of Present-Day American English.* Providence, 1967.

Latviešu valodas biežuma vardnica. (Častotnyj slovar'latyšskogo jazyka), tom II, gazety i žurnaly, c. I, Riga, 1968.

Muller, Ch., *Some Recent Contributions to Statistical Linguistics.*- Statistical Methods in Linguistics, Skriptor, Stockholm, 1976.

Mandelbrot, B., *Linguistique statistique macroscopique.* In: Apostel, L., Mandelbrot, B., Morf, A., *Logique et théorie de l'information.* Paris, 1957.

Minsky, M., *A Framework for Representing Knowledge.*-AI. Memo 306. MIT Artificial Intelligence Laboratory. Cambridge Mass., 1974.

Reid, J.R., *A French Word-Frequency Distribution Curve.*- Language, vol. 20, 1944, N 4.

Rinsland, H., *A Basic Writing Vocabulary of Elementary School Children.* New York 1945.

Sambor, J., *Badania statystyczne nad slownictwem (na materiale "Pana Tadeusza").* Kraków, 1969.

Schank, R.C., *Using Knowledge to Understand.*-Theoretical Issues in Natural Language Processing. Cambridge (Mass.), 1975.

Yule, G.U., *The Statistical Study of Literary Vocabulary.* Cambridge, 1944.

Vašak, P., *Průmyslový Strojový překlad - Tvórba.* Tydeník pro politiku vědu a kulturu. 7.XII, 1977, s. 2.

INDEX OF NAMES

INDEX OF SUBJECTS